AMERICAN SCIENCE SERIES, ELEMENTARY COURSE

THE HUMAN BODY

A BEGINNER'S TEXT-BOOK

OF

ANATOMY, PHYSIOLOGY AND HYGIENE

WITH DIRECTIONS FOR ILLUSTRATING IMPORTANT FACTS
OF MAN'S ANATOMY FROM THAT OF THE LOWER
ANIMALS, AND WITH SPECIAL REFERENCES TO
THE EFFECTS OF ALCOHOLIC AND OTHER
STIMULANTS, AND OF NARCOTICS

BY

H. NEWELL MARTIN, D.Sc., M.A., M.D., F.R.S.

Professor of Biology in the Johns Hopkins University

AND

HETTY CARY MARTIN

NEW EDITION REVISED

NEW YORK
HENRY HOLT AND COMPANY
1885

COPYRIGHT, 1884,
BY
HENRY HOLT & CO.

In Memoriam—
Irving Stringham

PREFACE.

THIS little book is an attempt to express accurately and yet in simple language, those facts concerning the structure and actions of the living human body which it is desirable, for practical purposes, that every one should know. It is essentially a school-book of personal hygiene. Little, if any, more Anatomy and Physiology is introduced than is necessary to make clear the reasons, as regards the preservation of health, for following or avoiding certain courses of conduct. This, of course, includes all the broad facts of Human Anatomy and Physiology; but subjects of merely professional importance or of purely scientific interest have been omitted. As regards Hygiene, attention is for the most part only directed to matters which are usually within the easy control of each individual. It seems useless to burden boys and girls with sanitary laws which need the aid of a physician or engineer for their successful application.

A very earnest attempt has been made to present the subject so that children may easily understand it, and, wherever possible, to start from familiar facts and gradually lead up to less obvious ones. In this part of the task, which was really the most difficult, I have had so much aid from my wife's experience in teaching young pupils, that her name properly has a place on the title-page.

We both desire to express our obligations to Miss Frances F. Bauman, who placed freely at our disposal the results of her long and eminently successful experience in teaching Physiology to children.

As appendices to certain of the chapters there are practical directions for the illustration of various facts in Anatomy and Physiology, which can be shown to pupils without any special apparatus, or any material not easily obtained.

Particular attention has been given to the action on the body of the more commonly abused stimulants and narcotics, especially alcohol.

H. NEWELL MARTIN.

JOHNS HOPKINS UNIVERSITY, June 30, 1884.

CONTENTS.

THE HUMAN BODY.

CHAPTER I.

THE GENERAL PLAN ON WHICH THE HUMAN BODY IS BUILT.

1. Why we should Learn about our Bodies.—Suppose you had given to you a delicate instrument, such as a watch: you would desire to be told something of the way it was made, how it was to be used, and what was apt to harm it. Even a little knowledge of these things would help you to take better care of the watch.

Now every one of us is responsible for the care of a *body* made up of many more parts than we find in a watch, and any of them liable to be injured in numberless different ways. If all the parts work well we are in *health*, able to enjoy our lives, do our work, and aid those who are less fortunate. If we lose our health we not only can do less and enjoy less ourselves, but are likely to become a burden upon others. It is therefore one of

1. If you have a watch, what ought you to know of it, and why? What is the nature of the machine given to every human being to take care of? If it is kept in good order, what is the result? If not? What then is our duty with regard to it?

our first duties to learn enough about our bodies to be able to avoid doing things likely to harm them, or neglecting to do that which is for their welfare.

2. What Anatomy is.—We could not look at the watch without seeing that it was made up of different pieces, as case, and face, and hands; and a glance at the works inside would show us that dozens of parts, such as wheels, and pivots, and springs, and screws, without which the portions we see on the outside would be useless, were fixed together in a special way to make the watch.

Likewise, on looking at the outside of the body you easily perceive head, and neck, and trunk, and arms, and legs; and if you could see into the inside you would find hundreds of other parts, which move the parts you see and make them useful. The science which teaches us the shape and size of all the parts of the body, where they are placed in it, and how they are joined together, is named *Human Anatomy.*

3. What Physiology is.—On examining the parts of which a watch is made we find that each has its use: the case to protect the works, the glass to let us see the face and yet keep out dust, the hands to show the hour, the spring to keep it going, and so forth.

In like way it is found that the various parts of the body have their uses: as the eyes to see, the mouth to eat, the legs to walk with. The science which teaches the uses of all the parts of the body, more particularly of its inner parts, is named *Human Physiology.*

2. What do we easily find out on examining a watch? In this respect how may the body be compared to a watch? What is Human Anatomy?
3. Why are there many parts in a watch? In the body? What is Human Physiology?

4. What Hygiene is.—Lastly, when you had learned
something of how the watch was made and what each part
of it had to do, you would know that certain things must
injure it; that it should be kept dry lest the steel springs
rust, and that the case must be kept closed to prevent
dust and grit from getting into the works. You might
also be told some things which it would take you a
longer time to find out for yourself; as, for example, that
if the watch is to be a good time-keeper it must be regularly wound up, and not at one time one day and at
another the next, or perhaps quite forgotten a third.

So, without learning much Anatomy and Physiology
you will readily see that certain things must be bad
for your body: such as getting wounds that will cause
great loss of blood, or going without food. The harmfulness of other things it might take you a long time to
find out by yourself; as, for example, that by breathing
foul air or taking too little sleep, eating imprudently or
drinking what is called "spirits," you might very easily
injure your body beyond cure. Unless you were warned
you would probably not discover the danger until too
late to avert it.

Just as a watchmaker could save you a great deal of
time and risk by giving the results of his experience as
to the best way to manage a watch, physicians and others
who have made a study of what is good and what bad
for the human body can save us much labor and danger
by telling what they have found out. The science which

4. Having examined a watch, what would at once occur to you about
its preservation? What studies teach you that certain things would
be bad for your body? Name some injurious habits that the experience of others warns you to avoid. What is meant by Hygiene?

teaches what is good and what hurtful to our bodies—in other words, how we may best preserve our health—is known as *Hygiene.*

5. **Organs and Functions.**—The separate parts of which the body is made up are called *organs:* thus the eye is the organ of sight, the teeth are organs of chewing, the stomach is an organ of digestion. The use of any organ is spoken of as its *function :* thus the function of the eye is seeing, of the ear hearing, of the hand grasping.

6. **The Structure of Organs.**—The human body, like a watch, not only has numerous parts, but these parts are made of different materials. Taking the hand, for example, we observe on the outside, skin, nails, and hairs. If the skin were removed we should see below it some *fat*, just like that in beef and mutton. Under the fat, in the ball of the thumb you would find some red flesh, called *muscle*, which answers to the lean of meat. Beneath all the rest would be white hard *bones.* At the finger-joints where the ends of separate bones come near together you would see covering each a thin layer of gristle or *cartilage.* And binding together the skin and fat and muscles and bones would be found a stringy substance which, as it unites all the rest, is called the *connective material.*

7. **Tissues.**—Each kind of material used in constructing the body is called a *tissue:* and each tissue has its own peculiar properties. *Connective tissue* is tough and suited

5. What is an organ ? Give examples. What is a function ? Illustrate.
6 What is meant by the structure of an organ ? Describe the structure of the hand.
7. What is a tissue ? Name and describe some tissues. Name some liquids of the body.

to bind parts together. *Bony tissue* is stiff and useful to support softer parts. *Cartilage tissue* is elastic and forms admirable springy cushions between the hard bones. *Muscle tissue* has power to move parts to which it is joined; and so on.

In addition to the solid tissues, liquids form part of the body: as the *blood* which we see flow from a cut finger, and the *saliva* which moistens the mouth.

8. The General Plan on which the Body is Built.—If a man's body were sawed in two down the middle, so as to separate it into right and left halves, we should see something like Fig. 1, if we looked at the cut surface of the right half. On examining the figure you see that there are two chief cavities or chambers in the body, having between them the row of bones *e e;* these bones together form the *back-bone* or *spine.* The chamber, *B, C,* in front of the back-bone is much the larger; it is named the *ventral cavity.* The other chamber, *a, a',* is the *dorsal cavity.*

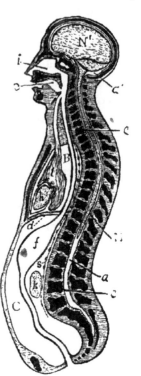

Fig. 1.—A section along the middle of head, neck, and trunk. *b,* the chest, and *c,* the abdominal division of the ventral cavity separated by the diaphragm, *d. a',* the enlarged upper end of the dorsal cavity, containing the brain, *N'. a,* the lower narrower part of the dorsal cavity containing the spinal cord, *N. e, e,* the partition formed by the back-bone between the ventral and dorsal cavities. *i,* the nose. *o,* the mouth. *l,* the lungs; the tube leading down to them is the windpipe. *h,* the heart. *f,* the stomach; the tube leading down to it is the gullet; the tube passing from the stomach to the lower end of the trunk is the intestine. *k,* a kidney. *s,* the sympathetic nervous system.

8. How does the plan on which a watch is made compare with that on which the body is constructed? In a human body cut down the middle what chief divisions would you find?

9. The Ventral Cavity, as you perceive in the fig-ure, does not reach up into the neck or head. It exists only in the trunk of the body, and is divided into an upper story, *B*, the *chest* or *thorax*, and a lower story, *C*, the *abdomen*, by a partition, *d*, which forms the floor of the thorax and the ceiling of the ab-domen. This partition is the *diaphragm*. How far in your own body the chest-cavity extends you can find out pretty accurately by beginning at the bottom of the neck and feeling down along the middle of the front of your trunk till you feel no more bones through the skin: that level marks the bottom of the thorax.

10. Contents of the Thorax.—On Fig. 1 you will also see that the mouth, *o*, and the nose, *i*, join behind, and that from the place of meeting two tubes run down the neck. The front one of these tubes is the *windpipe* or *trachea;* after entering the thorax it ends in the *lungs, l.* In the thorax is also placed the *heart, h.*

11. Contents of the Abdomen.—The second of the tubes above referred to is the *gullet* or *œsophagus.* It runs right on through the chest and diaphragm into the abdomen, and there opens into the *stomach, f.*

The air we take in when breathing goes along the windpipe to the lungs, but no further: the food and drink which we swallow take a longer road along the gullet to the stomach.

In addition to the stomach, the *liver,* the *intestines* or *bowels,* and the *kidneys, k,* lie in the abdomen.

9. Where is the ventral cavity? Name its divisions. The parti-tion. How can you trace the chest or thorax in your own body?
10. Name contents of thorax.
11. What is the course of the gullet? Its use? Use of the wind-pipe? Name the organs which lie in the abdomen.

12. What would be seen if the front of the Thorax and Abdomen were cut away.—This is represented in Fig. 2. Stretching across from side to side is seen the diaphragm,

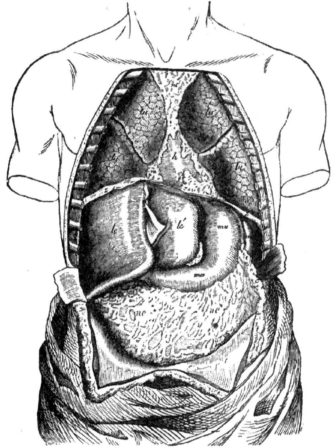

Fig. 2.—The trunk of the body opened from the front to expose the contents of the ventral cavity. *s s*, the diaphragm; *lu, lu'*, the lungs; *h*, the heart; *ma*, the stomach; *mi*, the spleen; *ne, ne*, the membrane (*great omentum*) which lies in front of the intestines and kidneys.

12. Name the parts which would be exposed if the front wall of chest and abdomen were cut away. State their positions.

z z. Above the diaphragm, in the thorax, are the *lungs,*
lu, lu. Between the lungs is the *heart, h,* partly covered
by fat and other things. Below the diaphragm is the
liver, le, le', the *stomach, ma,* and the *spleen, mi.* Hanging
down from the stomach is a sort of apron, *ne, ne;* if it
were lifted up we should find behind it the *intestines* and
the *kidneys.*

13. The Dorsal Cavity (*a, a'*, Fig. 1) is found in the head
and neck as well as in the trunk of the body. If the back
or top of a man's head were cut away the upper end of
the dorsal cavity would be opened and we should find it
to be a large chamber having the *brain, N',* in it. In the
neck and trunk the dorsal cavity is a narrow tube containing in its upper two-thirds the spinal cord, *N.*

14. Man is a Vertebrate Animal.—The presence of the
ventral and dorsal cavities with a hard partition between
them is a chief fact in the anatomy of the human body:
it shows that man is a *vertebrate animal,* that is to say, is
a *back-boned animal,* and belongs to the same great group
as fishes, reptiles, birds, and beasts. Worms, clams, and
insects are *invertebrate animals,* that is, have no back-bone.

15. Man's Place among Vertebrates.—We have seen
that man is a vertebrate, or back-boned animal. Though
all vertebrates are alike in the general plan of their
structure, there are such differences that zoologists divide them into classes. The most important of these

13. Where does the dorsal cavity lie? Name its contents and give
their position.
14. Why is man a vertebrate animal? Name some other vertebrates. How are vertebrates distinguished from invertebrates? Give
examples of invertebrate animals.
15. Why are vertebrates divided into classes? To which class
does man belong? Name some other mammalia. How do mammalia differ from other vertebrates?

classes is the *mammalia*, to which man belongs. Ordinary four-footed beasts, and monkeys, are also mammalia. The mammalia differ from all fishes, reptiles, and birds, first, in the possession of organs, the *mammary glands*, which provide milk for the young; second, in possessing *hair;* third, in having the chest separated from the abdomen by a *diaphragm.*

16. **The Intellect of Man** makes him superior to any other animal and supreme in the world.

His *power to form conceptions of right and wrong* and his *knowledge of moral responsibility* give him yet greater superiority. But as a *material object* only, do anatomists study man's body, and they therefore classify it among the bodies of other animals according as it differs from or resembles them in the arrangement of its parts.

17. **Chemistry of the Body.**—Suppose you put a green stick into the fire: what happens? At first it hisses and gives off steam; then it begins to burn; if you draw it out when half burned you find it a black mass of charcoal; if you put it back you find most of the charcoal will burn away, but some ashes will be left which you cannot make burn. If, instead of a green piece of wood, a man's body be burned, we find the same results. From this we learn (1) that the body contains water; (2) that it contains solid things which will burn; (3) that it contains solid matters, the ash, which will not burn.

16. What makes man superior to all other animals? What gives him yet greater superiority? From what standpoint is man studied by anatomists? How classified?

17. What would be the action of fire on green wood? On man's body? Hence what do we learn? What name is given the materials which burn up? What those which will not burn? Of what is every tissue composed? In which do we find most water? Mineral matter? Animal matter?

The things going to make up the body and capable of being burned are known as *animal matters;* the ashes are *mineral matters.* In every tissue of the body there are water, animal matter and mineral matter. In some a great deal of water, as in the blood; in others a great deal of mineral matter, as in the bones and teeth, which owe their hardness to lime; in still others a great deal of animal matter, as in fat and muscle; but everywhere some of all three.

18. Summary.—*Anatomy* is concerned with the form and structure of the parts of the body.

Physiology with the uses of the parts and the ways in which they work.

Hygiene with the conditions of life which promote the health of the body.

The *materials* of the body are hard or soft, solid or liquid, and are fitted for different purposes.

Tissue is the name given to each of the materials, whether blood, bone, muscle, fat, or any other.

The *organs* are formed of tissues combined in various ways. Each organ has its own particular duty, or *function*, which in health it performs in harmony with all the others.

Vertebrates are animals having back-bones—such as man, beasts, birds, reptiles, and fishes. Their bodies contain two main cavities, dorsal and ventral. In the dorsal cavity are the brain and spinal cord. The ventral cavity contains lungs, heart, stomach, liver, intestines, and kidneys.

18. What does anatomy deal with? Physiology? Hygiene? What have we learned of the materials of the body? Of tissue? Of the organs? Of vertebrates? Of invertebrates? Of mammalia? Of the chemical constituents of the body?

Invertebrates are animals having no back-bones—such as worms, clams, and insects.

Mammalia is the highest of the several classes of verte-brates and includes man, monkeys, and four-footed beasts. It is characterized by the presence of *mammary glands;* by the fact that the ventral cavity is separated into *chest* and *abdomen* by the *diaphragm;* and by having more or less of the surface covered with hair.

Water, animal matters and *mineral matters* compose the body. If it be burned the animal matters are consumed; the mineral matters remain in the form of ashes.

CHAPTER II.

THE SKELETON.

1. The Skeleton —By the skeleton of any animal we usually mean those hard parts which remain behind when the softer parts have decayed; as the shell of a clam or crab, or the bones of a bird or beast. In our own bodies, bones form the chief part of the skeleton; but other things help. A very young infant has a skeleton, but this skeleton is made for the most part of *cartilage*, or *gristle*, and not of bone. As the child grows, more and more bone takes the place of the cartilage; but even in old age some cartilage remains. Moreover, a skeleton consists not merely of all the bones of a body, but of all the bones united together in their proper places. In our bodies they are bound together by tough stringy *connective tissue*. The skeleton of the living body, as distinguished from a dead skeleton made of dry bones joined together by wires, is therefore made up of three different things; namely, bones, cartilages, and connective tissue.

2. The Bones, two hundred and six in number (see table, p. 22), form the hardest, and stiffest, and heaviest

---

1. What is a skeleton? What change takes place in the skeleton of a child as it grows? How are the bones of a skeleton put together? What are the materials of the living human skeleton?

2. Number of bones in the skeleton? What part of it do they make? How do they provide support? Protection? How concerned in movement?

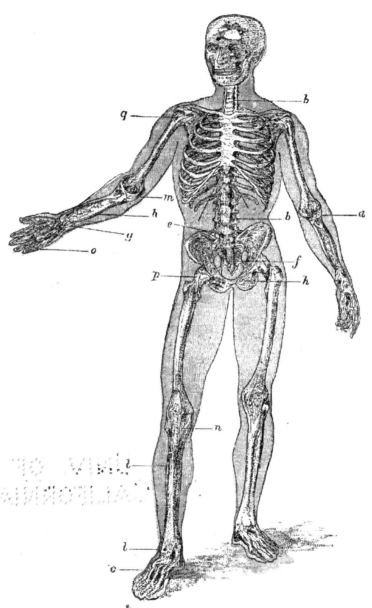

PLATE I.—THE BONES, JOINTS, AND LIGAMENTS.

EXPLANATION OF PLATE I.

A front view of a human skeleton with the ligaments and some of the cartilages in place.

For the names of the bones see the description of figure 3.

a Ligaments of the Elbow-Joint.
b The Ligament which is connected to the ventral surfaces of the bodies of the Vertebræ.
e Ligament connecting the Innominate Bone to the Spine
f Ligament connecting the Innominate Bone to the Sacrum.
g The Ligaments of the Wrist-Joint.
h The connective-tissue Membrane which fills up the interval between the two bones of the Forearm.
l A similar Membrane between the two bones of the Leg, and, lower down, *l*, ligaments of the Ankle-Joint.
k A connective-tissue Membrane which fills up a hole in the Innominate Bone.
n Ligaments of the Knee-Joint.
o o Ligaments of the Toes and Fingers.
p Capsular (bag-like) Ligament of the Hip Joint.
q Capsular Ligament of the Shoulder-Joint.

part of the skeleton. United in various ways, they provide a strong framework which supports the softer organs, and in some places, as the skull (Fig. 6) and thorax (Fig. 5), make strong boxes or cages in which delicate organs, such as the brain or lungs, lie safe. The bones are also concerned in the movements of the body; nearly all muscles pull first on some bone or other, and when the bone is made to move, it of course carries with it the surrounding soft parts.

3. Cartilage is what we know in meat as *gristle :* it is stiff enough to keep its shape, but can be bent with tolerable ease; it is also elastic, so that it springs back to its proper shape, like a piece of whalebone, as soon as the force which has bent it ceases to act. You can easily feel on your nose the difference between bone and cartilage. The skeleton of that part of it near the forehead is made of bone, and that of the lower part of cartilage. We can push the tip of the nose to either side, or up and down, but when we stop pressing, it returns to its place. The skeleton of that part of the ear which projects from the side of the head is also made of cartilage.

Cartilage is used in parts of the skeleton which have to be moderately stiff, but at the same time pliable and elastic.

4. Connective Tissue is used for several different purposes in the body. To understand this, let us imagine a quantity of very fine strands of silk, some twisted into

3. What is cartilage? Its properties? How used in the nose? In the ear? Throughout the skeleton?
4. To what may connective tissue be compared? Name and character of its threads? How are the cords made? The membranes? The loose portion? Where do we find networks of connective tissue? Give an example.

strong cord or rope, some woven into firm bands, some left in loose masses, and some made up into fine network. Connective tissue consists of threads, called *fibres*, which are much tougher and finer than any strand of silk. In some parts of the body these threads are united to form cords named *ligaments*, which bind bones together. Elsewhere the fibres are woven into bands or *membranes* which surround and support various parts. Lying in the crevices between different organs, forming a soft packing for them, we find loose fluffy bundles of connective tissue. Finally, very fine networks of this tissue run all through most of the organs, like the veins or ribs through the leaf of a plant, and support and unite their parts. If you watch the cook cut up a piece of suet, you will see the stringy connective tissue which penetrates it in all directions, and which must be removed from the fat because it will not melt in cooking.

5. **Action of Alcohol upon Connective Tissue** —All intoxicating liquors, such as wine, brandy, whiskey, beer, etc., contain alcohol and are known as alcoholic drinks. One very serious change in the body frequently produced by drinking such, is an excessive growth of the connective-tissue networks, especially in the liver and the kidneys. The tissue becoming too abundant crushes and slowly destroys the chief liver and kidney substance which it was meant to protect and support. The results are incurable diseases. (See pp. 132, 189.)

6. **The Bony Skeleton** (Fig. 3), like the body itself, may be described as consisting of head, neck, trunk, and

5. What are alcoholic drinks? How do they affect connective tissue?
6. Of what parts does the bony skeleton consist?

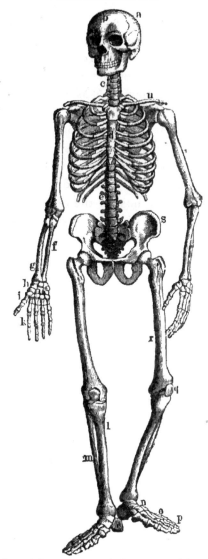

FIG. 3.—The bony skeleton. *a, b,* the skull; *c, e,* the back-bone; *d,* the breast-bone; *u,* the collar-bone; *t,* the humerus; *f,* the ulna; *g,* the radius; *h,* the carpal bones; *i,* the metacarpal bones; *k,* the phalanges of the fingers; *s,* the hip-bone; *r,* the thigh-bone; *q,* the knee-pan; *l,* the shin-bone or tibia; *m,* the fibula; *n,* the tarsal bones; *o,* the metatarsal bones; *p,* the phalanges of the toes.

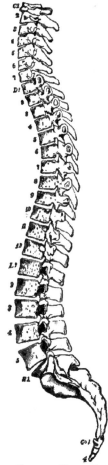

FIG. 4.—The spinal column viewed from the left side. *C*1-7, the vertebræ of the neck; *D*1-12, the vertebræ behind the thorax. *L*1-5, the vertebræ of the loins; *S*1 to *Co*1, the sacrum; *Co*1-4, the coccyx.

limbs. Its central part, which bears all the rest, is a stout, bony pillar, the *back-bone, c, e,* on the top of which is the skull.

7. The Back-Bone, Vertebral Column, or Spine, is represented in side view in Fig. 4. Its upper part is made of twenty-four short thick bones piled one upon another, and each called a *vertebra.* Between each pair of vertebræ there is placed during life an elastic cartilaginous cushion. The lower part of the spine consists of two bones; a large one, the *sacrum,* extending from *S*1 to *Co*1; and a much smaller, the *coccyx,* reaching from *Co*1 to the end.

Projecting from the back of each vertebra (to the right in the figure) is a bony bar, called its *spinous process.* Through the skin along the middle of the back we can feel the tips of these processes, and it is their presence which has given the name *spinal column* to the whole.

A canal runs through the whole back-bone except the coccyx, and opens into the skull-chamber above. It is the lower part of the *dorsal cavity* (*a,* Fig. 1), and, as we have already learned, contains the spinal cord.

7. What other names has the back-bone? Divisions of its upper part? Lower part? What is the spinous process? The dorsal cavity?

8. Uses of the Mode of Structure of the Spinal Column.—
The elastic cushions between the vertebræ make the
whole column springy and prevent the transmission of
sudden jars along it. By this means the soft brain, carried in the skull on its top, and the spinal cord lying in

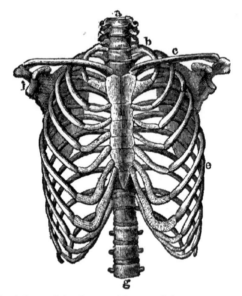

FIG. 5.—The skeleton of the thorax, with some of the vertebræ of the neck and
loins. *a*, lower neck vertebræ; *b*, the first rib; *c*, the collar-bone; *d*, third rib; *e*,
seventh rib; *g*, last loin-vertebra; *h*, the breast-bone; *i*, the shoulder-blade.

it, are protected from injury in running and jumping.
These cushions also allow of a little bending between
each pair of vertebræ, so that the spine as a whole may
be bent a good deal. But no sharp bend, such as would
nip the spinal cord, which lies inside it, can take place
at any one point.

8. Of what use is the cartilage between the vertebræ in running or
jumping? In bending?

9. The Ribs and Breast-Bone (Fig. 5).—The ribs are twenty-four slender curved bones, twelve on each side of the chest. Every rib is attached behind to a vertebra, the top one to the first vertebra below the neck. In front, each rib ends in two or three inches of cartilage. The *breast-bone* or *sternum*, *h,* lies in front of the chest. Attached to its sides are the cartilages of most of the ribs. The two lowest ribs are not joined to the breast-bone and are sometimes called the *free* or *floating ribs.*

10. The Skull (Fig. 6) is made up of twenty-nine bones (see table, p. 22); those behind and above arranged to form the brain-box; and those in front, to support the face.

The organs of four of our senses, viz., those of hearing, sight, smell, and taste, are also protected by the skull-bones.

11. The Sutures.—Except the lower jaw-bone, which is attached to the rest of the skull by a joint, to let us open and close our mouths, nearly all the skull-bones are very firmly united. In most cases the union is by a dovetailing, like that used by cabinet-makers. Each bone has its edge notched and fits accurately to the edge of the next. This sort of junction between bones is called a *suture.* It is well seen in Fig. 6 between the bone *Pr* and those in front of, behind, and below it.

12. How the Brain is Protected.—The dome-like form

9. What is the number and form of the ribs? How attached behind? How do they end in front? How attached to the breast-bone? Floating ribs?

10. How many bones in the skull? Use of those behind and above? Those in front? What other organs do they protect?

11. How is the lower jaw-bone attached? Union of other skull-bones? What is a suture?

12. What is the advantage of the dome-like form of the skull? Il-

of the crown of the head gives it great strength. This you will realize if you take an egg by its ends between finger and thumb, and try to crush it: you will find that

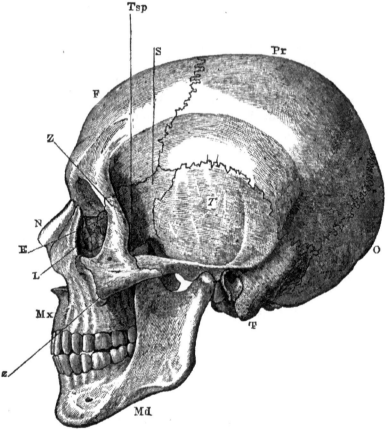

FIG. 6.—Side view of the skull. *Pr*, parietal bone; *O*, occipital bone; *T*, temporal bone; *S*, sphenoid bone; *F*, frontal bone; *Z*, malar, or cheek-bone; *N*, nasal bone; *E*, ethmoid bone; *L*, lachrymal bone; *Mx*, upper jaw-bone; *Md*, lower jaw-bone.

lustrate. Describe the outer layer of the bones on the sides and top of the brain. The next. The innermost. To what may this arrangement of the skull-bones be compared?

yõu cannot, although egg-shell is thin and brittle. The
bones on the sides and top of the brain-case are made
up of three layers: an outer, tough and fitted to bear
without breaking, blows from a heavy blunt object. Then
comes a much softer layer which deadens any jar that
might result from a blow on the head, and hinders its
transmission to the brain. Inside is a layer of very hard
bony matter, almost like glass, and admirably fitted to
stop or turn aside any pointed instrument which might
have penetrated the outer layers. If you turned upside-
down a thin china teacup, wrapped around it a covering of
raw cotton, and over this put a thin casing of tough wood,
anything placed under the cup would be protected from
blows, jars, and piercing, much as your brain is protected inside the skull.

13. **The Skeleton of each Upper Limb** contains thirty
bones and is attached to the trunk by the *shoulder-girdle.*

14 **The Shoulder-Girdle** presents on each side a *collar-
bone* or *clavicle,* in front (*u,* Fig. 3, and *c,* Fig. 5), and a
shoulder-blade or *scapula* (*i,* Fig. 5), behind. The collar-
bone and shoulder-blade unite near the shoulder-joint.

15. **The Bones of the Arm and Hand** (Fig. 3) are:
(1) the *arm-bone,* or *humerus, t,* which reaches from the
shoulder to the elbow; (2) two *forearm-bones* lying side
by side between the elbow and the wrist; the one on
the thumb-side is the *radius, g,* and that on the little-
finger side the *ulna, f;* (3) twenty-seven hand-bones.

13. How many bones in the forelimb? How is it attached to the
trunk?
14. What bone forms the front part of the shoulder-girdle? Behind? Where do these bones unite?
15. Name the bones of the arm. Give position of humerus.
Radius. Ulna. Carpal bones. Metacarpal. Phalanges.

Eight of the hand-bones are small and lie close to the wrist-joint: they are the *carpal bones*, *h*.   Five, the *meta-carpal bones*, *i*, lie in the palm of the hand; fourteen, the *phalanges*, *k*, are placed, three in each finger and two in the thumb.

16. The Skeleton of the Leg and Foot contains, like that of the arm and hand, thirty bones, and is attached to the side of the sacrum by the hip-bone.

17. The Hip-Bones (*s*, Fig. 3), one on each side, meet in front and form, with the sacrum, a bony ring enclosing the lower part of the cavity of the abdomen or belly. This ring is named the *pelvis*.

18. The Bones of the Leg and Foot are: (1) the *thigh-bone*, or *femur*, *r*, reaching from the hip-joint to the knee: it is the longest bone in the body; (2) the *tibia* or *shin-bone*, *l*, and *fibula*, *m*, running side by side from knee to ankle-joint; (3) the *knee-pan* or *patella*, *q*, in front of the knee-joint; (4) twenty-seven foot-bones.

Seven of the foot-bones, named *tarsal bones*, *n*, lie below the ankle-joint and support the heel; five *metatarsal bones*, *o*, follow these; and fourteen *phalanges*, *p*, are found in the toes, two in the great toe and three in each of the others.

16. How many bones in the leg?   How attached to the sacrum?
17. Describe the hip-bones.
18. Name the leg-bones.   State position of femur.   Tibia.   Fibula. Patella.   Tarsal bones.   Metatarsal bones.   Phalanges.

TABLE OF THE SKELETON.

The Bony Skeleton: 206 bones.

Head, Neck, and Trunk: 80 bones.
Skull: 29 bones.
Brain-case, 8 bones, namely:
Occipital bone, at back of head.......................... 1
Frontal bone, in forehead............................... 1
Parietal bones, on top and sides of head................ 2
Temporal bones, in the temples......................... 2
Sphenoid bone, on floor and sides of brain-box. 1
Ethmoid bone, between top of nose and brain case...... 1
— 8

Face-bones, 14, namely:
Lower jaw-bone... 1
Vomer, between the nostrils............................ 1
Upper jaw-bones.. 2
Palate-bones, supporting part of the roof of the mouth.... 2
Malar bones, supporting the cheek below and outside
the eye 2
Lachrymal bones, between nose and eye-socket......... 2
Nasal bones, on roof and sides of nose................. 2
Turbinate bones, inside the nose....................... 2
— 14

Ear-bones, 6, three on each side, within the ear, namely:
Malleus, or hammer-bone....................... 1
Incus, or anvil-bone 1
Stapes, or stirrup-bone 1
$3 \times 2 =$ 6

Hyoid bone, to which the root of the tongue is attached....... 1
29

Vertebral Column: 26 bones, namely:
Cervical (neck) vertebræ............................ 7
Dorsal vertebræ, at back of thorax..................12
Lumbar (loin) vertebræ............................. 5
Sacrum... 1
Coccyx... 1
— 26
Ribs: 24 bones, on each side twelve... 24
Sternum (breast-bone) 1
——
80

LIMBS AND THE BONES UNITING THEM TO THE TRUNK: 126 bones.
Shoulder-girdle: 4 bones, on each side two, namely:

Clavicle, or *collar-bone*... 1
Scapula, or *shoulder-blade*......................... 1
                                                          ——
                                                 $2 \times 2 = 4$

Arms: 60 bones, on each side thirty, namely:

Humerus........ 1
Ulna..................... 1
Radius........................ 1
Carpal or wrist bones......................... 8
Metacarpal bones... 5
Phalanges........ 14
                                              ——
                                   $30 \times 2 = 60$

Hip-bones: on each side one...................... 2
Legs: 60 bones, on each side thirty, namely:

Femur, or thigh-bone........................... 1
Patella, or knee-pan........................... 1
Tibia, or shin-bone 1
Fibula, or "small bone of the leg"........... ... 1
Tarsal (ankle and heel) bones....... 7
Metatarsal bones.............................. 5
Phalanges...,.................................14
                                              ——
                                   $30 \times 2 = 60$
                                              ——
                                           126

CHAPTER III

THE STRUCTURE, COMPOSITION, AND HYGIENE OF THE BONY SKELETON.

1. The Parts of the Humerus —Though bones differ in shape and size, we may get a pretty good idea of the way they are all built by studying the humerus, Fig. 7. This presents a central rounded portion, or *shaft*, bearing at each end an enlargement, the *articular extremity*. The shaft lies between the dotted lines *x* and *z*. One use of these large ends is to give more room for the fastening on of muscles.

2. Internal Structure.—If the humerus be sawed in two lengthwise (Fig. 8) we find that its shaft is hollow; the space is the *marrow cavity*, *a*, and during life is filled with a kind of fat. We also see that there are two kinds of bony substance; one is hard and close, the other loose and spongy. The hard bone, *b*, lies on the outside, and is thick in the shaft; it forms only a thin layer in the extremities, which are filled with spongy bone, *c*. The large marrow-cavity does not extend into the extremities.

3. Why Bones are Hollow.—All bones either contain a marrow-cavity or are filled up with loose spongy tissue.

1. Describe the humerus as viewed on its outside. For what are its large ends useful?

2. What would we find inside the shaft? The extremities?

3. What do all bones contain? Why are they not filled with hard bone? Why are the iron pillars used in building made hollow?

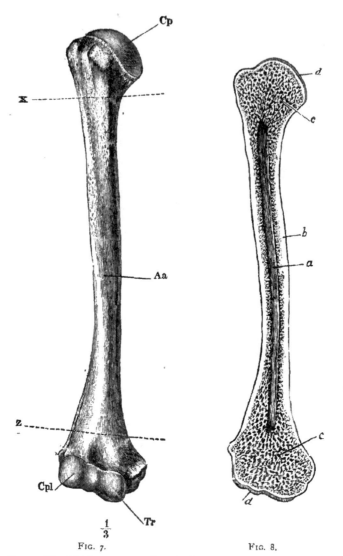

FIG. 7.

FIG. 8.

FIG. 7.—The right humerus, seen from the front.
FIG. 8.—The humerus cut open. *a*, marrow-cavity; *b*, hard bone; *c*, spongy bone; *d*, cartilage.

If they were of hard bony substance throughout, they would be either very heavy and unnecessarily strong, or else too slender to give surface enough for the attachment of muscles and other organs. A given quantity of material if arranged in the form of a tube will bear a much greater weight than if it were made into a solid rod of the same length as the tube. For this reason, iron pillars used in buildings to support ceilings and floors, are hollow. To cast them solid would make them much heavier without great increase of strength.

4. **How Bones are Nourished.**—When the humerus is in the body, it is closely surrounded by a connective-tissue membrane, the *periosteum*. This membrane is full of blood which nourishes the bone by means of innumerable little channels passing into and branching all through it. These channels are too small to be seen without a microscope, but even the most close-grained part of every bone is full of them. As long as the humerus is growing thicker, the periosteum is making new bone on its inner side. If this membrane is peeled off, the bone dies. The parts of the articular extremities (*Cp, Tr, Cpl,* Fig. 7) which meet other bones at the shoulder and elbow-joints are covered by cartilage instead of periosteum.

5. **The Chemical Composition of Bone.**—The dried bone of a man in middle life, consists of two parts of mineral to one part of animal matter. The minerals give the bone its hardness and stiffness; they may be obtained separate

4 With what is the humerus surrounded? How does the periosteum nourish the bone? What happens if it be peeled off? Where is cartilage found instead of periosteum?
5. Of what does the dried bone of a middle aged man consist?

by thoroughly burning a bone. The animal matter may be obtained by soaking a bone for a few days in an acid which dissolves away the minerals.

The mineral matter by itself has still the form of the bone, but is very brittle The animal matter by itself also has the form of the bone, but is soft and easily bent. The two mixed together, as they are in the skeleton, make our bones hard enough to support the rest of the body, and tough enough not to be easily broken. The animal matter also makes the bones tolerably flexible and elastic: some savages make their bows from the ribs of large animals.

In childhood the animal matter, and in old age the mineral matter, of bone is more abundaht than in middle life. Therefore the bones of an old person are brittle and easily broken, while those of a child often bend when the bones of an adult would break.

6. Gelatin.—When a bone is boiled in water for several hours, most of its animal matter is turned into *gelatin*, and dissolved in the water. Gelatin is a useful food; most of that which we buy for making jelly is made from bones. For soup we use bones as well as meat, and by long boiling extract the gelatin from them. In a piece of meat as ordinarily cooked most of the gelatin remains in the bones, which are therefore useful for soup and should not be thrown away.

What is the use of the minerals ? How may they be obtained separate? How the animal matter? Characters of mineral matter? Of the animal? Use of having both in a bone ? At what time of life is the animal matter most abundant ? Why are an old person's bones easily broken ?

6 How may we get gelatin from a bone? Why are bones left from a piece of meat useful in making soup ?

7. Hygiene of the Bony Skeleton.—Except hair and teeth, bones are the parts of the dead body which most resist decay. Nevertheless living bone is readily altered in shape, especially in young persons, by continued or frequently repeated pressure or strain. This is well illustrated by the curious forms which some nations give to their skulls (Fig. 9) by tying boards or bandages on the heads of their children.

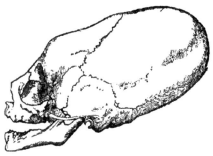

FIG. 9.—Skull of a child of the tribe of Chinook Indians (inhabiting the neighborhood of the Columbia River), distorted by tight bandaging so as to assume the shape considered elegant and fashionable by the tribe.

8. Why Children should have their Feet Supported and should Sit Straight.—The bones of a child being rich in the softer animal matter are tolerably flexible, and may be readily made to grow out of shape. Therefore children should never be kept sitting on a bench so high that the feet are not supported. If this precaution be neglected the thigh-bones become bent over the edge of the seat by the weight of the rest of the limb and may be made crooked for life.

7. What parts of the dead body decay most slowly? How may living bone be altered in form? Illustrate.
8. Why should the feet of children be supported when sitting? Why is it important to sit straight? Why should children not be encouraged to walk too soon?

For the same reason it is important to sit square and straight at the table when writing or drawing, and with the shoulders level: otherwise the spinal column may become curved to one side.

Young children should not be encouraged to walk too early, lest they grow bow-legged, their leg-bones not being stiff enough to bear the weight of the upper part of the body.

9. Usefulness of the Arch of the Instep.—The bones of the foot (Fig. 10) are arranged to make a springy arch which

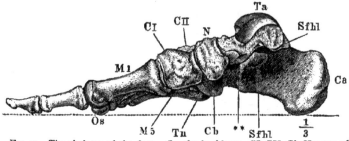

FIG. 10.—The skeleton of the foot. *Ca*, the heel-bone; *CI, CII, Cb, N*, some of the tarsal bones; *Os*, the front end of the metatarsal bones; *Ta*, the surface which makes the ankle-joint with the tibia and fibula, and bears the weight of the body in standing and walking; *M1*, metatarsal bone of the great toe.

rests on the ground by the heel-bone, *Ca*, behind, and by the front ends, *Os*, of the metatarsal bones in front. On the crown of the arch is the surface, *Ta*, where the foot joins the leg at the ankle-joint. At this joint the weight of the body is borne. The many small bones in the arch glide over one another a little when the crown of the arch is pressed upon; but spring back into place when the pressure is removed. This elastic arch of the foot

9. Describe the construction of the instep. Why is it arched and elastic? To what may it be compared? How may we learn something of the jarring saved us by the instep? Illustrate the usefulness of a well-arched instep in prolonged walking.

lessens the jarring which would be transmitted to the
spinal column, and thence to all the rest of the body,
were the foot flat or rigid.   A well-arched instep is
therefore rightly considered beautiful; it makes the step
easier and more elastic.

We may compare it to a carriage-spring, which gives
a gentle sway to the vehicle and prevents sudden jolting.
How much jarring the instep saves us, may be readily
learned by walking across a room on the heels.   For a
steady, even, long-continued tramp, like that of a police-
man, a foot well-arched under the instep is of great importance: it not only saves the upper parts of the body
from injury, but much diminishes the fatigue of walking.
Men who desire to join the police force but who are
" flat-footed," are rejected; experience having proved
that such persons cannot walk the daily " rounds."

10. Why High-heeled Boots are Hurtful.—When we
walk on the heels, we are jarred at each step because the
arch of the instep is not used as a spring.   If we walk on
the toes, this is not the case, as the elastic front half of
the foot is brought into action.   But walking or running
on the toes is fatiguing because it demands extra muscular effort.   Boots with high heels lead practically to walking on the toes.   The sole of the boot forms such a
slope, high behind and low in front, that the whole foot
slides forward on it, and the heel has no place on which
it can bear firmly and take its share of the work.   The
arch of the instep is made useless, and the toes slip along

10. Why are we not as much jarred when we walk on our toes as
if we walk on our heels ?   Why is walking on the toes fatiguing?
What are the consequences of wearing high-heeled boots ?   Of boots
with pointed toes ?

until they are squeezed into the toe of the boot; and on them all the weight of the body is there carried. The so-called "French heel," placed right under the arch of the instep, makes that piece of Nature's mechanism perfectly useless.

The results are an awkward, ungraceful gait; and undue fatigue, leading to omission of proper healthy exercise, to the loss of many innocent pleasures, and often to

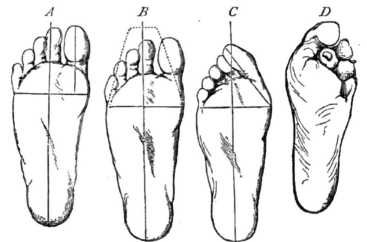

FIG. 11.—*A*, natural form of the sole of the foot; *B.* the same with the outline of an ordinary fashionable boot; *C, D,* feet which have been made to grow out of form by wearing such a boot.

neglect of duties whose performance necessitates walking.

Continued wearing of narrow-toed boots, especially if they have also high heels, leads to permanent distortion of the foot. Its front part being forced into the toe of the boot by the weight of the body, the toes are pushed out of place, frequently pressed over one another (Fig. 11), and made useless; while corns and bunions are

developed, making the walk still more painful and less
graceful.

11. The Evils of Tight Lacing can only be properly understood after we have studied the use and working of the
heart and lungs (Chap. XVI.). With our hands we can
press in our lower ribs and narrow the chest-cavity; but

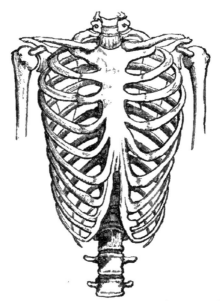

FIG. 12.—Skeleton of the chest of a woman, twenty-three years of age, deformed
by tight lacing. Compare with the natural skeleton, Fig. 5.

as soon as we cease the pressure, the ribs spring back
to their place. If, however, a tight corset be worn for
weeks or months, the ribs gradually yield to it and
change their shape. The result is a deformed chest-
skeleton (Fig. 12). The lower ribs press on the liver,

11. How does tight lacing alter the ribs? What organs are injured
in consequence?

injuring it; and the bottom of the chest-cavity is so narrowed that the heart and lungs are cramped for room.

12. What should be Done when a Bone is Broken.—When a bone is broken, it is said to be *fractured.* The muscles on each side of the break are very apt to pull the pieces of the bone out of place. Therefore the broken bone needs to be *set* into place, and then held by splints and bandages so that the ends be kept together until they unite. To set a broken bone, often needs great skill and a thorough knowledge of. anatomy. A medical man should be summoned without delay, as the parts around the fracture usually swell very rapidly, making the exact position of the break hard to find out, and the replacement of the pieces of the bone more difficult Until skilled aid arrives, the sufferer should be kept as quiet as possible: cloths dipped in cold water and frequently renewed may be applied to keep down swelling and inflammation

13. How a Broken Bone is Knit together again.—A watery liquid first collects between and around the broken ends. This gradually thickens, becoming jelly-like, and then of the hardness of gristle, though it does not become actual cartilage. It is chiefly made by the periosteum, which becomes very active where the bone is broken, and makes this uniting material in such abundance that it forms quite a thick ring all round the fracture. This ring, named the *callus*, is afterwards

12. What is a fracture? Why does a broken bone require to be set? Why should a doctor be summoned at once? What should be done until he arrives?

13. What first happens when Nature begins to repair a broken bone? Next? What makes this first uniting material? What is the callus? Its use? What finally becomes of it? What takes place inside the callus?

hardened by lime being deposited in it. It forms a sort
of natural splint, and strengthens the bone until the
ends have firmly grown together.   Then it is slowly
absorbed, and after a few months hardly a trace of
it is left. The callus may be compared to the metal
band which is used to hold together the two parts of
a broken umbrella-handle. Inside the callus, new bone
slowly forms in the gristly layers between the broken
ends, and unites them. The surgeon usually removes
his artificial splints when the callus has become well
developed.

CHAPTER IV.

THE ORGANS OF MOVEMENT : MUSCLES AND JOINTS.

1. Articulations.—Wherever two bones meet in the body an *articulation* is formed. In some articulations the bones are fixed immovably together, as in the sutures of the skull, (p. 18); in others, to enable us to move, the ends of the bones are so shaped and so fastened together that one can slide over the other. Articulations of this kind are called *joints*. Joints may be compared to hinges between bones: examples are found between the lower jaw-bone and the rest of the skull; at shoulder, elbow, wrist, hip, knee, ankle; and between the bones of the fingers and toes.

2. The Movements of the Body are brought about by soft red organs named *muscles*. The lean of meat is muscle, so every one knows what dead muscle is like. Living muscle has the power of shortening, or *contracting*, with great force. When a muscle contracts it pulls its ends together and swells out in the middle; in other words, *it becomes shorter and thicker*. If you watch the front of your forearm while you forcibly bend your wrist, you can observe, through the skin, the muscles becoming shorter and thicker. Nearly always the two ends of a

1. What is an articulation? Of what kind of articulation are the sutures of the skull examples? What is a joint? Name some joints.
2. What is the use of muscles? What is dead muscle like? What power has living muscle? How does it change its shape in contracting? Illustrate. To what are the ends of a muscle usually fixed? What results when a muscle contracts?

muscle are attached to separate bones, between which
a joint is placed; and when the muscle contracts it produces movement at the joint. The joints and muscles
are the chief organs of movement.

3. Joints.—As an example of a joint we may take
that at the hip (Fig. 13).

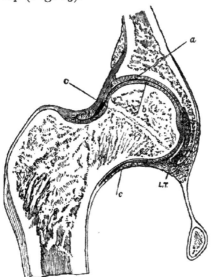

Fig. 13.—The hip-joint, sawed through its middle. The rounded head of the
thigh-bone is seen to fit into the cup or socket of the hip-bone. *a*, cartilage lining socket; *b*, cartilage covering end of femur; *c, c*, capsular ligament; *L T*, round
ligament.

On the outer side of the hip-bone (*s*, Fig. 3) is a cup-
like hollow which receives the round upper end of the
thigh-bone. Lining the cup is a thin layer of cartilage,
and covering the end of the thigh-bone is another. The
cartilage is extremely smooth and is kept moist by a few
drops of *joint-oil*, or *synovial liquid*, so that the end of the

3. Describe the hip-joint. What use is the cartilage? The synovial fluid? The ligaments?

femur rolls very easily in the hollow, or *socket* The cartilage forms a yielding cushion which hinders the bones from scratching or chipping one another.

To keep the bones in place and prevent too free movement, strong bands of connective tissue, called *ligaments*, unite them, being fixed above to the hip-bone and below to the femur. Many powerful muscles also pass from one bone to the other, and keep them pressed close together.

4. Ball-and-Socket Joints.—A joint like that at the hip, where the round end of one bone fits into a cavity in which it can roll in any direction, is called a *ball-and-socket joint.* It allows more free movement than any other kind. At the shoulder there is another ball-and-socket joint.

5. Hinge-Joints —In hinge-joints the ends of the bones are not evenly rounded on all sides, but one bone has projecting ridges which slide in grooves on the other. The result is that the only movements possible are to and fro, or in one direction and back again, like a door on its hinges.

The knee is a hinge-joint: it can only be bent and straightened; or, as physiologists say, *flexed* and *extended.* Between the phalanges of the fingers there are other hinge-joints.

6. Pivot-Joints.—In pivot-joints one bone rolls round another.

A good example is the joint which permits us to turn the head from side to side.

The uppermost vertebra (Fig. 14), which carries the

4. What is a ball and-socket joint?
5 Describe a hinge-joint. Examples.
6. What are pivot-joints? Describe the atlas. What is the odon-

skull, has been fancifully named the *atlas vertebra*, after
the fabled giant of antiquity who was believed to bear
the heavens on his shoulders.   It is ringlike in form and
the space which it surrounds is separated by a ligament,
L, into a smaller front and larger back division. . In the
larger division the spinal cord lies. Into the smaller projects a bony peg (*D*, Figs. 14 and 15), called from its
shape the *toothlike* or *odontoid process*, which springs from

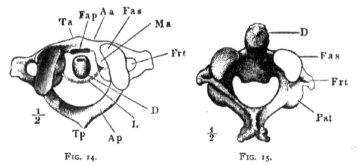

FIG. 14.   FIG. 15.

FIG. 14.—The atlas vertebra seen from above. FIG. 15.—The axis vertebra,
L, the ligament which divides the space surrounded by the atlas into a back
portion, containing the spinal cord; and a front portion, containing the odontoid
process, *D*, of the axis, round which the atlas rolls when we turn the head to either
side.

the second or *axis vertebra*. Knobs on the under side of
the skull fit into the hollows (*Fas*, Fig. 14) on the atlas:
when we turn the face to right or left the atlas, carrying
the skull with it, rolls around the odontoid process.

Another kind of pivot-joint is found in the forearm.
Lay the hand and forearm flat on a table, palm upwards.
Without moving the shoulder-joint at all, it will be easy

toid process ? What happens at the joint between atlas and axis
when we turn the face to one side ? Where is there another kind of
pivot-joint ? What is the position of radius and ulna when the palm
of the hand is turned up ? When turned down ?

EXPLANATION OF PLATE II.

A view of the muscles situated on the front surface of the body, seen in their natural position. It must be undeistood that beneath these muscles many others are situated, which cannot be represented in the figure.

Muscles of the Face, Head, and Neck:

1. Muscle of the Forehead. This, together with a muscle at the back of the head, has the power of moving the scalp
2. Muscle that closes the Eyelids. The muscle that raises the upper eyelid so as to open the eye, is situated within the orbit, and consequently cannot be seen in this figure.
3, 4, 5. Muscles that raise the Upper Lip and angle of the Mouth.
6, 7. Muscles that depress the Lower Lip and angle of the Mouth By the action of the muscles which raise the upper lip, and those that depress the lower lip, the lips are separated.
8 Muscle that draws the Lips together, so as to close the Mouth.
9 Muscle of the Temple (Temporal Muscle).
10 Masseter Muscle 9 and 10 are the two chief muscles of mastication, for when they contract, the movable lower jaw is elevated, so as to crush the food between the teeth in the upper and lower jaws
11. Muscle that compresses the Nostril. Close to its outer side is a small muscle that dilates the nostril
12. Muscle that wrinkles the Skin of the Neck, and assists in depressing the lower jaw.
13 Muscle that assists in steadying the Head, and also in moving it from side to side
14. Muscles that depress the Windpipe and Organ of Voice. The muscles that elevate the same parts are placed beneath the lower jaw, and cannot be seen in the figure.

Muscles that connect the upper extremity to the trunk. Portions of four of these muscles aie iepresented in the figure, viz.:

15 Muscle that elevates the Shoulder. Trapezius Muscle.
17. Great Muscle of the Chest, which draws the Arm in front of the Chest (Great Pectoral Muscle).
18. Broad Muscle of the Back, which draws the Arm downwards across the back of the Body (Latissimus Dorsi).
19. Serrated Muscle extends between the Ribs and Shoulder-blade, and draws the shoulder forwards and rotates it, a movement which takes place in the elevation of the arm above the head (Serratus magnus).

At the lower part of the trunk, on each side, may be seen the large muscle which, from the oblique direction of its fibres, is called,

20. Outer Oblique Muscle of the Abdomen.

Several muscles lie beneath it. The outline of one of these,

21. Straight Muscle of the Abdomen, may be seen beneath the expanded tendon of insertion of the oblique muscle. These abdominal muscles, by their contraction, possess the power of compressing the contents of the abdomen.

Muscles of the upper extremity:

16. Muscle that elevates the Arm (Deltoid Muscle).
22. Biceps or Two-headed Muscle (see also page 41).
23. Anterior Muscle of the Arm   This and the Biceps are for the purpose of bending the Fore Arm
24 Triceps, or Three-headed Muscle.   This counteracts the last two muscles, for it extends the Fore-arm
25. Muscles that bend the Wrist and Fingers, and pronate the Fore-arm and Hand—that is, turn the Hand with the palm downwards. They are called the Flexor and Pronator Muscles.
26. Muscles that extend the Wrist and Fingers, and supinate the Fore-arm and Hand—that is, turn the Hand with its palm upwards. They are called the Extensor and Supinator Muscles.
27. Muscles that constitute the ball of the Thumb. They move it in different directions
28. Muscles that move the Little Finger.

Muscles which connect the lower extremity to the pelvic bone. Several are represented in the figure

29 Muscle usually stated to have the power of crossing one Leg over the other, hence called the Tailor's Muscle, or Sartorius, its real action is to assist in bending the knee.
30. Muscles that draw the Thighs together (Adductor Muscles).
31. Muscles that extend or straighten the Leg (Extensor Muscles)   The muscles that bend the leg are placed on the back of the thigh, so that they cannot be seen in the figure.

Muscles of the leg and foot:

32. Muscles that bend the Foot upon the Leg, and extend the Toes
33. Muscles that raise the Heel—these form the prominence of the calf of the Leg
34. Muscles that turn the Foot outwards
35. A band of membrane which retains in position the tendons which pass from the leg to the foot
36. A short muscle which extends the Toes

The muscles which turn the foot inwards, so as to counteract the last named muscles, lie beneath the great muscles of the calf, which consequently conceal them. The foot possesses numerous muscles, which act upon the toes, so as to move them about in various directions. These are principally placed on the sole of the foot, so that they cannot be seen in the figure. Only one muscle, 36, which assists in extending the toes, is placed on the back of the foot.

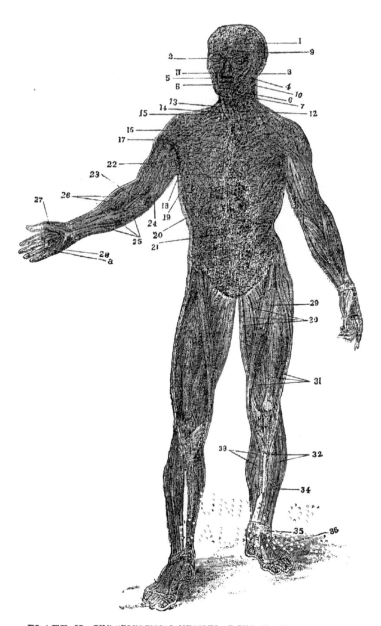

PLATE II.— THE SUPERFICIAL MUSCLES OF THE FRONT OF THE BODY.

to turn the hand, palm downwards. This is done (Fig. 16) by rolling the lower end of the radius, which carries the hand, around the ulna. When the palm is upward the radius and ulna lie side by side as shown at *A*; while it is being turned downward, the lower end of the radius rolls around the ulna and at last crosses it to get on its inner side, as shown at *B*.

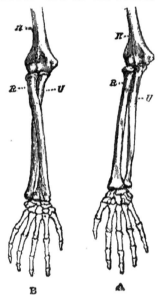

7. The Muscles of the human body are more than five hundred in number. They vary in size from tiny ones inside the ear, not half an inch long, to that (29, Plate II.) which passes from the pelvis to the tibia and is eighteen inches or more in length. All muscles have the power of shortening and thus of pulling other parts (usually bones) to which their ends may be attached. After a muscle has shortened and done its work, it lengthens again, or *relaxes*. In addition

FIG. 16.—Bones of the forearm and hand. *A*, the palm turned forwards or upwards (*supination*), and the radius and ulna parallel; *B*, the palm turned downwards or backwards (*pronation*), and the radius crossing the ulna.

to their chief function of moving the body, muscles clothe the skeleton and make the form round and shapely; they aid in enclosing cavities, as the mouth and abdomen; and they help to hold bones together at joints.

7. How many muscles in the body? Their size? On what do our movements depend? What other functions have muscles?

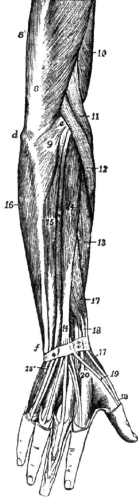

FIG. 17.—The muscles on the back of the hand, forearm, and lower half of the arm, as exposed on dissecting away the skin.

8. The Parts of a Muscle.—In its most common form, a muscle consists of a red soft middle part, called its *belly*, which tapers towards each end and passes into very tough white cords named *tendons* or *sinews*. The tendons may be compared to ropes, tying the working part of the muscle, namely its belly, to the bones which the muscle has to move. The hard cord-like tendons of the muscles which bend the fingers, can easily be felt through the skin in front of the wrist.

9. The Muscles of the Arm, some of which are shown in Fig. 17, may be taken to illustrate the structure and arrangement of nearly all muscles. We see that some (8, 11, 12) pass over the elbow-joint from arm to forearm. Others (14, 15, 16, 17, 18) start from the ulna or radius and pass over the wrist-joint to the hand. Near the wrist most of them end in slender tendons, which are kept in place by a strong cross-band of connective tissue (✝✝). The skin has been dissected away from the back

8. What parts has a muscle? Their uses?
9. Describe the course of some of the arm-muscles.

of the middle finger to show the ending of tendons on it.

10. How we may Recognize the Working of a Muscle.—The shortening of a muscle, when it is at work, is sufficiently shown by the way it pulls the bones; as when we bend the elbow-joint or the fingers. The thickening may be seen and felt on the *biceps-muscle* (Fig. 18), in front of the humerus, when the elbow is bent; or on the muscles of the ball of the thumb, when we move the

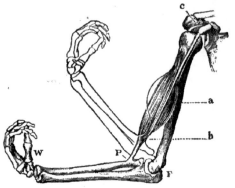

FIG. 18.—The biceps-muscle and the arm-bones, to illustrate how the elbow-joint is bent, when the biceps-muscle contracts and becomes shorter and thicker.

thumb so as to make it touch the little finger. When a muscle contracts, its belly becomes harder. The swelling and hardening of a contracted muscle are daily illustrated when a school-boy bends his elbow as powerfully as he can and then invites another to feel his "biceps."

11. Muscles not directly attached to the Skeleton.—Most of these surround openings, which they close when they contract. Thus around the mouth-aperture is a ring

10. How may we recognize the shortening of a working muscle? The thickening? The hardening?

11. Give examples of muscles not directly attached to bones.

of muscle (*orbicularis oris*, 8, Pl. II.) which shuts the mouth,
or if more vigorously contracted purses out the lips, as
when a child holds up its mouth to be kissed. A similar
ring-like muscle (*orbicularis palpebrarum*, 2, Pl II) surrounds the opening between the eyelids and closes the
eyes.

12. How the Muscles are Controlled.—It is very clear
that we could not do what we wanted to do if our muscles contracted at random: they must be held in control;
kept at rest when their action is not needed, and made to
work when it is. If the muscles closing the mouth contracted when we tried to put food into it we should be
in a bad plight. All the muscles are directed and guided
in their work by the nervous system (Chap. XVIII.). From
the brain and spinal cord *nerves* run to them, governing
all and making them work together in harmony ; those
which straighten the elbow-joint are not, for example,
permitted to pull when we desire to bend it. In *convulsions,* the controlling nervous organs cease their guidance;
the muscles contract in all sorts of irregular and useless
ways; and, often, since those which˘ produce exactly
opposite movements contract at the same moment, the
whole body is made stiff.

13. Involuntary Muscles.—The muscles hitherto considered are all more or less under the control of our will
We can make them contract or prevent their contraction as
we choose. They are called the *voluntary muscles.* There
are other muscles whose working we cannot control;

12. What power must we have over our muscles ? What is the use
of the nerves of the muscles ? In what organs do they commence ?
How do the muscles behave during a fit of convulsions ? Why?
13. What are voluntary muscles ? Involuntary ? Where found ?
Use ?

they are named *involuntary muscles.* Involuntary muscles are not attached to the skeleton nor concerned in our ordinary movements, but lie in the walls of hollow organs, as the stomach and intestine (Chap. XI.). When they contract they push along the contents of these organs.

14. As a general rule all the movements most necessary for keeping the body alive, as those which cause the blood

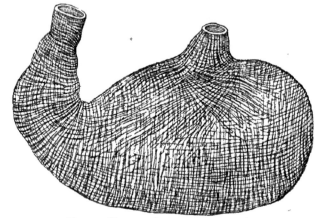

Fig. 19.—The muscular coat of the stomach.

to flow to all organs or food to travel along the alimentary canal, are taken by Nature out of our control, and performed by involuntary muscles. It is, however, impossible to draw a sharp line between voluntary and involuntary muscles. The breathing muscles are partly subject to our will: any one can draw a long breath when he chooses. But in ordinary quiet breathing, we are quite unconscious of the working of these

14. What class of movements is not subject to our will? Illustrate. What is said concerning the breathing muscles? Give instances in which other voluntary muscles contract against our will.

muscles; and even when we pay heed to it, our control
is limited: no one can hold his breath long enough to
suffocate himself. Any one of the voluntary muscles
may be thrown into activity, independently of or even
against the will, as we see in the "fidgets" of nervous·
ness. When we call any muscle voluntary, we mean that
it may be controlled by the will, but not that it neces-
sarily always is so.

15. Standing.—There are two reasons why a young
infant cannot stand: the first is that its skeleton is not
firm enough to bear its weight; the second is that it cannot guide and manage its muscles. After the bones are
strong enough a child has still to *learn* to stand. We
all at last become by practice able to do so without
thinking about it; but standing always demands that a
great many muscles shall be at work, and be guided by
the brain. The part the brain takes, although we usually know nothing about it, is shown by the fall which
results from a violent knock on the head. This may
break no bone and injure no muscle, and yet the man who
has received it falls stunned and helpless to the ground.
His brain has been so shaken that it ceases for a time
to do its work, and the consequence is that the muscles,
released from control, cease to do their work; so until
the brain recovers, the man cannot stand.

**16. How our Brains come to Control the Muscles without
our being Conscious of it** —A child learning to stand has
to take a great deal of trouble; it has to think about
what it is doing all the time. After a while, it gives less

15. Why cannot an infant stand? How is the brain concerned in
standing? In what way is this fact shown?
16. Give an example of an action once performed with trouble

and less thought to the proper action of the muscles' of standing; and at last its brain does the work without any thinking about it at all. The child then stands, as it breathes, almost or quite unconsciously. This is a very curious and a very important fact. It is but one example of many, showing that actions of our muscles which once cost thought and effort, come at last to be done without either. Practice not only " makes perfect," it also makes easy that which before was difficult. The trouble with which we learn to ride or swim, or strike the proper keys of a piano, thinking about every necessary movement, and the ease with which we come at last to do these things, are other examples of the same fact. When any muscular action which was at first performed with difficulty and by " willing" to do it, comes to be performed almost unconsciously, without our will, we say a *habit* has been formed. When the brain and muscles have been trained to work together in this unconscious way, it is as hard or harder to break the habit than it was to acquire it. A practised rider would have to take a good deal of trouble to fall off his horse under ordinary circumstances, or a good swimmer to drown himself.

This tendency of the brain and muscles to do at last without the will, or against it, that which they have often done before in consequence of the will, is of the greatest importance. It is the physiological reason for acquiring good habits and avoiding bad. The more often we do anything, the easier it is to do it again, and the harder to avoid doing it.

which at last comes to be done unconsciously. Other illustrations. What is a habit? How do habits come to control us? What is the physiological reason for forming good habits and avoiding bad?

17. The Muscles concerned in Standing.—In consequence of the flexibility of the ankle-, knee-, and hip-joints, a

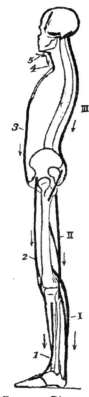

dead body cannot be balanced on its feet, as a marble statue may be. When a man stands, the joints would bend, were they not braced and held firm by muscles. When we stand, muscles (Fig. 20, 1) in front of the ankle-joint, and others (*I*) behind it, contracting at the same time, keep that joint from yielding. In the same way, muscles (2) in front of the knee- and hip-joints, are opposed by others (*II*) behind them, and when we stand, both contract and keep those joints rigid. The muscles (*III*) which run from the pelvis to the back of the head, in like manner pull against others (3 and 4) which run from the pelvis to the lower end of the breast-bone, and from the upper end of the breast-bone to the front part of the skull; their balanced contraction keeps the head erect. If one falls asleep while sitting or standing, the chin drops, because the muscles holding the head upright have relaxed their vigilance, and its front part is heavier than its back. Since the degree to which each muscle contracts when we

Fig. 20. — Diagram illustrating the muscles (drawn in thick black lines) which pass before and behind the joints, and by their balanced activity keep the joints rigid and the body erect.

17. How do the muscles enable us to stand? Why does the head fall forward if one goes to sleep standing? Why does it take time to learn to stand?

stand, must be exactly equal to the contraction of its antagonist on the opposite side of the joint, we easily see why it takes some time to learn to stand.

APPENDIX TO CHAPTER IV.

Many of the facts described in this chapter can be exhibited to a class with little trouble or expense.

1. The sutures may be well seen on the skull of a rabbit or sheep. All that is necessary is to boil it thoroughly and then pick the bones clean, and wash out the brain.

2. The structure of joints is easily exhibited on the fresh foot of a sheep or calf. On cutting open the joints the tough ligaments around them will be seen. The slippery synovial liquid covering the inside of the joint can be felt by the finger. The smooth gristle will be found to form a layer over the bones within the joint. A thin slice of it may be readily cut off with a knife, and its translucency, flexibility, and springiness exhibited.

3. An example of a ball-and-socket joint may be easily obtained by cleaning the thigh- and hip-bones of a rabbit or chicken.

4. For a good example of a hinge-joint the most easily available object is the skull of a dead cat. In this animal the lower jaw forms a perfect hinge-joint with the rest of the skull.

5. The pivot-joint between atlas and axis can be demonstrated on the bones of a sheep's head and neck, after the piece of meat has done its duty at table. For this purpose buy *mutton.* The odontoid process of a lamb is apt to separate.

6. The form and structure of muscles can readily be exhibited on the hind leg of a frog. Place the animal for a few minutes in a covered jar containing a pint of water to which has been added a teaspoonful of ether. When the creature has become quite unconscious take it out, cut off its head, and run a stout pin down its back-bone to destroy the spinal cord. In this way all chance of giving pain is avoided. Then divide the skin at the top of each leg and pull it off. Point out especially the muscles between knee- and ankle-joints, and their long white tendons, many of them running to the toes.

The leg of a chicken or turkey also affords an excellent object for examining tendons The bellies of most of the muscles which move the toes lie in the part of the leg known as the "drumstick." Their tendons run down the shank, and, if the skin be dissected off this, are readily found. Pulling some of the tendons bends the toes, pulling others straightens them, just as when they were pulled during life by the contracting bellies of the muscles in the drumstick

7. The nerve of a muscle can be easily shown on the calf-muscle of a frog's leg Cut the tendon (*tendo Achillis*) which attaches this muscle to the heel. Then turn the muscle up, so as to expose its under side. Its nerve will be seen, as a slender white thread, entering its deeper side a little way below the knee.

CHAPTER V.

CARE OF THE JOINTS AND MUSCLES.

1. Dislocations and Sprains.—When we slip or stumble, some joint has to share with the bones the strain of our effort to recover our balance; or the weight of the body if we fall. Accidents to the joints are accordingly quite frequent, and it is important to know how to manage them until medical aid can be obtained. A *sprain* is an injury in which the ends of the bones remain in place but the ligaments are stretched or twisted or torn. In a *dislocation*, the ligaments of the joint are torn, and the ends of the bones forced out of their proper positions.

2. How to Treat a Sprain.—The most important point is to give the joint complete rest. The injured ligaments become swollen and painful, and movement makes them worse. In the case of sprains of the finger and wrist the inflammation is often slight, and can be controlled by wrapping the joint in a moderately tight bandage for a few days, and keeping the arm in a sling so as to hinder it from being used. If the pain and swelling are great, the bandage should be kept wet with cold water. Sprains

1. Why are accidents to the joints frequent? What is a sprain? A dislocation?

2. What does a sprained joint most require? Why? How may a slight sprain of a finger or wrist be treated? What should be done when a knee or ankle is sprained?

of the knee and ankle joints are apt to be more serious, and if neglected or unwisely managed may lead to permanent lameness. In such accidents it is best to send at once for a surgeon; until he arrives, if the pain is great, apply cloths wrung out of hot water.

3. What to do in Case of a Dislocation.—The ligaments and soft parts around dislocated joints swell rapidly, and make it not only difficult to find out in what direction the bones have been displaced, but, after finding this, difficult to replace them. When a dislocation is suspected, get skilled advice as soon as possible; meantime keep the joint at rest. More harm than good is almost certain to be done by the twisting and pulling and pushing of persons ignorant of anatomy.

A dislocated finger may, however, be in most cases safely *reduced*—that is, have the bones put into place—by almost any one. What is needed is a strong pull, combined with pressure near the joint. The reduction of a dislocated thumb should never be attempted except by a surgeon.

4. Gout is a disease in which some joints, usually of the toes or fingers, become red, swollen, painful, and very tender. Gritty matter also accumulates in them, making the cartilage rough and the joint stiff. In nine cases out of ten gout is due to indolent and luxurious habits, too little exercise, too much animal food, and, above all, indulgence in alcoholic drinks. The disease, like many others produced by alcohol, tends to be inherited, and so some persons suffer from gout through

3. Why should a surgeon be called at once in case of most dislocations? How may a dislocated finger be usually reduced?
4. What is gout? To what often due? What is said concerning hereditary gout? Is gout ever fatal?

the fault of a parent; overwork may bring on an attack in such. Even those born with a gouty tendency may, however, usually escape if careful in their habits.

Gout is not merely painful but dangerous. It often attacks the heart, causing sudden death.

5. Rheumatism is a name given to different diseases. In *rheumatism of the joints*, or *rheumatic fever*, the ligaments of most of the joints of the body are swollen and inflamed. The inflammation often attacks also the membrane which covers the heart, or the valves inside it (Chap. XIII.), sometimes leaving incurable heart-disease when the rheumatism itself has gone.

The most common cause of rheumatic fever is prolonged exposure of the skin, especially if it be hot and perspiring, to chilling while the body is at rest. Therefore, when warm, especially avoid sitting in a draught. Exposure to cold air when exercising, or a plunge into cold water for a few minutes' swim, will not cause the disease; but sitting still in a current of air or in wet clothes, or sleeping in damp sheets, is apt to do so.

It is also well to know that rheumatic fever is more common, and more apt to cause heart-disease, in young persons than in old.

Chronic or *permanent rheumatism* may attack either the joints or the muscles. It makes the joints stiff, painful, and finally useless. The most frequent form of chronic rheumatism of the muscles is *lumbago*, in which the *lumbar* muscles in the lower part of the back are affected. Exposure to cold and wet is its most common cause; but

5. How are the joints affected in rheumatic fever? The heart? How is this disease commonly produced? Why specially dangerous to the young? What parts does chronic rheumatism attack? Its effects on the joints? What is lumbago? Usual causes?

the tendency to acquire it is much promoted by indulgence in alcoholic drinks.

6. The Importance of keeping our Muscles in Good Working Condition.—Man's mind is more than his body, but the mind turns its thoughts into deeds by means of the voluntary muscles. The better their state, the more promptly do they obey its commands; while a feeble or sluggish set of muscles will often bring to naught the best-laid plans of the mind.

Mind without the power of directing movement, would be a source rather of pain than pleasure. Muscles unguided by mind would make but a piece of machinery, as incapable of enjoyment as a steam-engine. Between these extremes, there lies a combination of vigorous well-trained brain and healthy active muscle, which is the highest condition of bodily welfare.

7. Hygiene of the Muscles.—Every time a muscle is worked, some of its substance is used up and turned into waste matters. Nourishment must therefore be brought to the muscle, that new substance be formed instead of that destroyed; and the waste matters, which would poison the muscle if they were allowed to collect, must be removed. Both of these things are done by the blood: and the blood must be kept in good condition by nourishing food and pure air, if the muscles are to be healthy and vigorous. No article of dress should press tightly on any muscle; if it does it will hinder a free flow of blood in it and interfere with its proper nourishment.

6 Why would our minds be of little use without our muscles? What is the highest condition of bodily welfare?

7. What happens to some of its substance when a muscle is used? What follows from this? What part does the blood play in keeping the muscles in health? What are necessary to keep the blood in proper condition? How may a tight garment injure our muscles?

8. Exercise.—After good air and food the most important condition for keeping the muscles healthy is that they be used regularly, or *exercised.* A muscle left in idleness dwindles in size and becomes worse in quality: instead of being hard, firm, and ready to contract, it becomes soft, flabby, and feeble. This fact is well shown in the muscles of an arm or leg which has been kept motionless for a few weeks while a broken bone is healing. When the bandages and splints are taken off, the muscles are nearly powerless, and much smaller than those of the opposite limb, which have been kept in use. Only by careful continued exercise, do they regain their former size and strength. The opposite fact, that muscles when used grow bigger and become more powerful, is illustrated by the huge " brawny" arm of a blacksmith.

9. Too Much Exercise is as Harmful as too Little.—When a muscle is at work, it is used up faster than new muscle-substance is made; also, waste substances are produced faster than the blood can carry them off. After a time, this causes a feeling of being tired, which is Nature's signal that it is time to rest. To exercise until we are a little tired, does no harm; indeed, rather benefits than hurts the muscles, if followed by proper repose. During a time of rest following moderate work, more blood than usual flows to the muscle, conveying more nourishment than is needed for its repair; and so it grows larger and

8. After good food and air what is next in importance for our muscles? How does continued idleness affect them? Illustrate. Give an example of the improvement of muscles by exercise.

9. Why do we feel tired after hard muscular work? What happens when we rest our muscles after moderate fatigue? How does repeated overwork affect the muscles? The body in general? What is necessary for healthy muscles? How is this illustrated by the heart? The breathing muscles?

stronger. Frequent exercise carried on to the point of great fatigue, leads to wasting away and weakness of the muscles as surely as does continued idleness. It also enfeebles the whole body and makes it more liable to many diseases.

Action and repose in turn, and neither in excess, are the conditions necessary for healthy muscles. In those whose action we cannot control by the will, we find this illustrated. The heart is a muscle which contracts seventy times or more every minute, in its work of pumping the blood all over the body. Yet the heart beats on year after year and feels no fatigue. The secret of this is that after every contraction it rests before it makes the next one. The muscles which cause the movements of breathing, teach us the same lesson. If they stopped their work for five minutes, we should die for want of fresh air in our lungs. After each breath we draw, they take their rest, and so keep at work fifteen or sixteen times a minute all life long.

10. The Proper Amount of Exercise is not the same for all persons. A strong healthy boy or girl runs about until pretty thoroughly tired, then goes home, eats a good supper, sleeps soundly, and wakes in the morning feeling all the better for the exercise. One who is delicate, should always rest as soon as the least fatigue is felt. Being delicate means, in most cases, that the organs of the body, the muscles along with the rest, only nourish themselves slowly; short exercise and long rest are therefore necessary. If a person who is not strong becomes

10. When should a delicate person stop exercising? Why? What is the result if a delicate person overexerts himself? How may healthful games be made injurious? What is about sufficient regular exercise for a healthy adult of sedentary habits?

greatly tired, he has little appetite, sleeps badly, and next morning still feels weary. His exercise, being more than his body is suited to bear, has done him harm. Many children (not to mention grown-up folks, who might be supposed to know better) run about in the excitement of some game, without realizing the fatigue, until after they have greatly overworked and injured their muscles.

A walk of from six to seven miles daily is about the proper amount of exercise for a grown person of ordinary strength, whose business is such as to keep him sitting most of the day and who is not able to take any other outdoor exercise. Horseback-riding is better for those who can afford it (p. 57).

11. When to Exercise.—Severe muscular work causes, as we have learned, great muscular waste, and demands an abundant supply of nourishment for the repair of the muscles. For this reason, violent exercise should not be taken after a long fast. Strong vigorous young people may walk several miles before breakfast and not suffer in consequence, but others had better wait until after eating, before undertaking any great muscular exertion. Neither should exercise be taken immediately after a meal. At that time, a great deal of blood is needed in the stomach and intestines to help in digesting the food (Chap. XI.); and it cannot be drained off to supply the muscles as it is during exercise, without risk of an attack of indigestion.

12. Exercise should be Regular.—When we work our voluntary muscles, we give the heart and lungs more work to do. The heart has to pump more blood to the muscles,

11. Why is it not wise to undertake hard muscular work when fasting? Just after eating?
12. How does muscular exercise affect the heart and lungs? What

and the lungs have to get rid of the extra waste matters (Chap. XV.). You know that after running fifty yards at full speed, you find yourself breathing faster and your heart beating quicker. If you are used to such racing, you soon get your breath again, and your heart quiets down; for those organs, having been gradually trained to work just as your muscles need their help, do it easily and comfortably. But if a boy who is not used to running starts off on a fast race, he soon has to stop, panting for breath, feeling his overstrained heart thumping in his chest, and probably with "a stitch in his side." Exercise leading to such results does harm. A healthy boy usually gets all right again in half an hour or so; but quite often fatal disease of the heart has been caused, even in strong young persons, by prolonged violent exercise to which they were not accustomed. Girls have in several cases died in consequence of excessive exercise with the skipping-rope Running to catch a train has often produced serious and lasting heart-disease in those who were weak or no longer young, and who were unused to such muscular exertion.

An occasional long walk at a moderate rate, or leisurely rowing a boat for an hour or two, if followed by a good rest, will not injure any one in ordinary fair health: those whose pursuits confine them to a desk most days are usually benefited by such exercise once a week. But fast running, or foot-ball playing, or rowing a race, should never be undertaken by those who have not gradually educated their bodies to bear violent exercise.

may you notice after running? What happens if a boy undertakes violent exercise without training? What organ is apt to be especially injured by unusual muscular exertion? Why is it better to miss a train than race to catch it if you are not used to running?

13. Proper Exercise Benefits not only the Muscles but the whole Body.—Suitable exercise makes the heart do more work in pumping blood over the body, but not enough extra work to injure the heart itself; the consequence is a better nourishment of all the organs. Such exercise also makes us breathe faster and deeper and so bring more air into our lungs. If the air be pure and fresh, this also benefits all the organs. The muscles take their toll of the general beneficial results; but if their work is not excessive, a good deal of the profit is left for other organs. The digestive organs are put in better working state, appetite is increased and more food eaten and used; the skin and other organs concerned in getting rid of wastes, work better; the brain is better nourished; the mind clearer; and work which without exercise was laborious and wearisome becomes easy and agreeable.

It is on these benefits to the body in general, which result from proper exercise of the muscles, that the duty of taking such exercise mainly rests. It is not a particularly lofty ambition to be strong enough to knock down another man in a stand-up fight, though there may be occasions when such muscular strength is very desirable. In the long-run, the world is guided and ruled by vigorous minds more than by muscular bodies. Exercise, in promoting the general health of the body, promotes mental vigor; and when pursued not for its own sake or for mere athletic glory, tends to quicken the intellect, invigorate the will, and strengthen character. Other things being equal, the healthy man or woman is the best in all the circumstances of life.

13. How does proper muscular exercise benefit the whole body? What is the chief reason which makes it a duty to take proper exercise?

14. Varieties of Exercise.—In walking, the muscles chiefly employed are those of the lower limbs and trunk; the muscles of the arms are hardly used. Rowing and riding are better, since in them nearly all the muscles are exercised. No one exercise employs in equal proportion all the muscles, and gymnasia, in which different feats of agility are practised so as to call different muscles into action as may be desirable, have a deserved popularity. It should be borne in mind, however, that in the arms delicacy of movement is more important to many persons than great strength. The fact that gymnastics are usually practised indoors is also a great drawback to their value. Out-of-door exercise in good or even moderate weather, is better than any other, and every one can at least take a walk. The daily " constitutional " is, however, very apt to become wearisome, and exercise loses half its value if unattended with feelings of mental re-.laxation and pleasure. Active games, for this reason, have a great value for young and healthy persons; lawn-tennis, base-ball, and cricket are all attended with pleasurable excitement, and are excellent also as exercising many muscles.

15. We cannot profitably Work Hard with both Brain and Muscle.—Few persons can continue to put both body and mind to severe daily work without risk. Many a college student has completely broken down his health in the attempt. Every one should, however, regularly use both

14. What muscles are left unexercised in walking? Why are rowing and riding better exercises? Why are gymnasia useful? What is the chief drawback to gymnastics? Why are active games especially valuable?

15. Why is it unwise for most persons to attempt to excel in both athletics and study? What should every one do?

mind and muscle, doing his *work* with one and simply exercising the other.   Thus both are kept in health.

16. The Action of Alcoholic Drinks on the Muscles.—Indulgence in beers, wines, or spirits never does any good to the muscular system of a healthy person, and often does great harm.   The injury may be direct or indirect; when indirect it is due to weakened will, impaired digestion, enfeebled heart, or disease of organs whose function it is to carry waste matters away from the body. How alcohol leads to these consequences we shall study in following chapters.

The action of alcohol on the power of the muscles has often been carefully studied.   Experiments prove that it is less on days in which spirits are taken (Chap. IX.).

Continued indulgence in alcoholic drinks causes change for the worse in the structure of the muscles.   The connective tissue and fat in them become too abundant and take the place of the proper muscular substance.   This consequence is especially frequent in the muscular tissue of the heart (p. 162).

16. How may alcoholic drinks indirectly harm the muscular system?   What has been observed as to their direct action on muscular power?   What changes in the structure of muscle are produced by continued alcoholic indulgence?   In what organ are they most frequently observed?

CHAPTER VI.

THE SKIN.

1. The Skin is the tough pliable membrane which surrounds and protects the rest of the body. It is not tightly fixed to the parts beneath it, but can glide over them or be pinched up in a fold; as you may easily observe on the back of your hand. The loose tissue which attaches the skin to parts under it contains a good deal of fat, except in very thin people; thus the form is made rounder and more beautiful than it would be if the skin fitted close to every bone or muscle beneath. This fat also serves as a soft padding or cushion protecting the deeper parts from injury by blows; and it checks loss of heat from the internal organs, by forming a sort of blanket around the body. In old age most of the fat is apt to disappear; the skin then falls into wrinkles, because it is too loose to fit neatly; and extra clothing is required to keep in the heat of the body.

2. Structure of the Skin.—The skin is made of two very different layers. The inner layer is named the *dermis*, and the outer the *epidermis* or *cuticle.* When your hand or foot becomes blistered in consequence of some exer-

1. What is the nature of the skin ? How is it attached to parts beneath ? Point out some uses of the fat under the skin. Why are old people wrinkled ?
2. What two layers compose the skin ? How is a blister produced ?

cise to which it is not accustomed, liquid collects between the cuticle and the dermis, causing the blister.

3. What we may Learn from a Blistered Hand.—When you open a blister, you feel no pain when cutting through its outer covering; but if you touch the raw surface beneath, it smarts. This shows that the epidermis has little or no feeling, while the dermis is very sensitive. You may also observe that when you cut through the cuticle, there is no bleeding; but if you gently prick with a pin-point the dermis under the blister, blood will flow. This shows that the dermis contains blood and the epidermis does not.

4. Other Illustrations of the Difference between the Dermis and Epidermis.—Without waiting for a blister, you may readily learn the facts described in the preceding paragraph. Take a small needle threaded with fine silk, and, if you are careful not to go deep, you can embroider a pattern on your Hand without causing pain or drawing blood. But if the point of the needle enters the dermis, you feel the prick, and a drop of blood is very likely to flow from the wound.

5. How the Epidermis is Shed and Renewed.—If you have ever seen an old brick house, you may have noticed that the bricks on the outside of the wall are worn away, crumbly, easily broken, and the mortar between them loose; while the bricks and mortar which lie deeper in the wall and have not been exposed to the weather, are perfectly sound. The epidermis (Fig. 21) is made up of millions

3. How may we learn from a blister which layer of the skin is sensitive? How discover which contains blood?

4. How may we in another way observe the same facts?

5. What might you notice on an old brick house? How do its walls resemble the epidermis? Of what is the epidermis made up?

of little pieces, called *cells*, joined together by a sort of glue. The cells may be compared to the bricks, and the

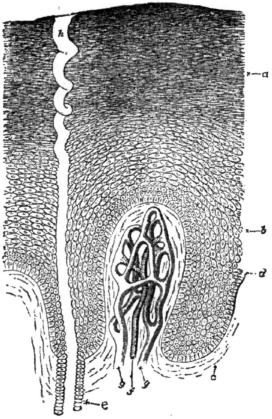

FIG. 21.—A thin slice of epidermis, greatly magnified. *a*, the outer or horny layer of the epidermis, made of old dry cells; *b*, the deeper moist layer of the epidermis, made of living growing cells; *d*, the deepest row of epidermic cells, lying next the dermis; *c*, the uppermost layer of the dermis; (it is seen to be elevated to form a papilla in which is a tuft of tubes, *f, g*, containing blood;) *h*, the duct of a sweat-gland.

glue to the mortar, of a wall. Each cell is so small that a powerful microscope would be needed to see one by itself, but, cemented together in thousands, they make up the tough epidermis, as we see it covering a blister. The cells, *a*, near the surface, exposed to the air and to wear and tear from rubbing against the clothes and other things, become different from the deeper cells, *b*. The outside part of the epidermis is in fact dead, and is being constantly shed and got rid of. Sometimes many cells come off together, as seen in the "peeling" of the skin after an attack of measles, or in the rolls of matter which a rough towel rubs off after a warm bath. These outer cells make what is known as the *horny layer* of the epidermis. It may be compared to a very thin sheet of india-rubber covering the body.

6. The deeper cells of the epidermis are kept moist and well nourished by a colorless liquid which exudes from the blood-vessels, *f, g*, of the dermis beneath them. They grow and divide, and thus make new cells, which in turn are pushed to the outside to build the horny layer. Beneath a blister, some of the deepest epidermic cells remain sticking on the dermis. Being well nourished, they multiply very fast and soon restore the whole thickness of the cuticle, so that in two or three weeks no trace of the blister remains.

7. **The Complexion** is due to the color of the deepest cells of the epidermis. In persons of blond or fair complexion, these cells contain very little dark coloring

How is the outermost part of the epidermis worn away and renewed? What does a rough towel rub off the skin after a bath? To what may the outer layer of the epidermis be compared?

6 Describe the life-history of the deeper cells of the epidermis. How is the epidermis restored after it has been removed by a blister?

7. How does the epidermis of a blond differ from that of a brunette? How does sunlight affect the epidermis? Why is the scar of a deep wound white?

matter. In those of brunette or dark complexion, this pigment is more abundant. In negroes there is a large amount of it.

Exposure to the air and to sunlight increases the quantity of coloring matter in the epidermis. Hence the skin darkens or "tans." If the whole thickness of the epidermis is destroyed, by a burn or wound, the deepest cells of the new epidermis do not usually form any coloring matter; therefore scars remain white, even in negroes.

8. Redness of the Skin and blushing are due not to changes in the epidermis, but in the dermis, which becomes fuller of blood. The red blood is then seen through the epidermis. Constant *pallor* or great whiteness of the skin, is a sign that there is not enough blood flowing in the dermis; it is usually an indication of disease. Some persons are pale from infancy and nevertheless healthy; but they are exceptional.

9. The Dermis consists of a close network of connective tissue, containing in its meshes many nerves, and numerous tubes filled with blood, named *blood-vessels.* It is the nerves (Chap. XVIII.) which give it feeling. When hides are tanned, the dermis is turned into leather. Its outer surface, next the epidermis, is not smooth, but presents numerous tiny projections, named *papillæ.* In Fig. 21 is shown a papilla containing a knot of blood-vessels. Other papillæ contain nerves instead of blood-vessels, and are concerned in the feeling of touch. On the palm of the hand, the papillæ of the dermis are

8. What is the cause of redness of the skin? Of pallor? Is constant pallor always a sign of disease?
9. Of what is the dermis composed? What are its papillæ? What do different papillæ contain? How are the fine ridges and furrows of the palm produced? What causes the deep lines of the palm?

arranged in rows. The epidermis fills up the hollows between those of the same row, but dips down between neighboring rows. In this way the fine ridges and furrows of the palm are produced. The deeper grooves of the palm, from whose size and course gypsies pretend to tell the fortune, have a different cause. They mark lines where the skin is most frequently folded in the movements of the hand, and where it is more tightly tied down to the parts beneath it.

10. Nails are made by a great development of the horny layer of the epidermis on the ends of the toes and fingers. This layer at these places becomes very thick, and grows out beyond the rest of the skin to form the edge of the nail. Our nails provide an armor to protect the tips of the tender fingers and toes, which are liable to many accidents. This protective use of the nail is well seen in the *hoof* of a horse or cow, which is but a thick nail. In beasts of prey, as cats and lions, the nails take the form of *claws* and are used in climbing and in catching prey.

Each nail is nourished by the dermis beneath it, and at its root. If it be torn off, or be shed in consequence of a blow, it is reproduced, provided the dermis also has not been seriously injured.

11. Hairs, long or short, coarse or fine, scanty or numerous, are found all over the skin except in a few places, as the palms of the hands and the soles of the

10. Of what are nails made? Use of the nails to man? What is a hoof? A claw? How are nails nourished? What is necessary in order that a " cast" nail may be replaced?

11. On what parts of the skin are there no hairs? What is a hair? What is the use of its papilla? What is the follicle of a hair? The root? What are the uses of hairs?

feet. A hair is a thread of epidermis which grows on the top of a papilla of the dermis (*i*, Fig. 22) placed at the bottom of a tiny pit in the skin. On the papilla, new epidermic cells are produced as long as the hair continues to grow. When a hair is shed or pulled out, a new one grows in its place if the papilla has not been injured. The part of a hair which lies within its pit or *follicle*, is known as its *root*.

In many of the lower animals, hairs have an important use as clothing. In man the hair of the head may serve this purpose to some extent; it also aids in protecting the skull from injury. The eyelashes keep dust from falling into the eye; and the fine hairs over most of the surface drag on their roots when pushed and aid in the sense of touch. The papillæ on which the hairs grow, are richly supplied with nerves.

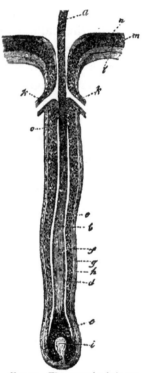

Fig. 22.—The root of a hair, embedded in the pit of the skin in which it grew. *a*, stem of the hair, cut short; *o, b*, root of the hair; *c*, swollen end of the root which fits on the papilla of the dermis, *i*, which nourishes the hair; *k, k*, openings of the ducts of oil-glands.

12. The Glands of the Skin.—Certain hollow organs of the body have as their special duty the preparation of

12. What are glands? A secretion? A duct? What glands pour their secretion on the skin?

peculiar liquids which they pass out through tubes. Such organs are called *glands :* the liquid which a gland collects or manufactures, is known as its *secretion;* and the tube through which the secretion is poured is named

a *duct.* Two kinds of glands empty their secretion on the skin. They are the *sweat,* or *sudoriparous, glands,* and the *oil,* or *sebaceous, glands.*

13. **The Sweat-Glands** (Fig. 23) make the perspiration. They are very slender tubes which reach from the surface, through epidermis and dermis, to the fatty tissue beneath the skin. There the tube coils up into a knot. These glands are found all over the skin, but not equally distributed. In the palm of the hand there are nearly three thousand to each square inch: on the middle of the back, where they are fewest, about four hundred in the same space. There are in all about two and a half millions of sweat-glands.

Fig. 23 —A sweat-gland. *a,* horny layer of the cuticle; *c,* deeper layer of the cuticle; *b,* the dermis. The duct of the gland is seen to run through the epidermis and dermis and end in a coiled mass placed in the fatty tissue beneath the skin.

14. **The Perspiration or Sweat** is a transparent colorless liquid, consisting mainly of water. Anything tending to heat the body causes perspiration to flow more freely, so its amount is very variable. On an average, it is at least two pounds daily. The sweat may dry up, or *evaporate,* as fast as it

13. Describe a sweat-gland. What is the number of these glands in a square inch of the palm? Of the skin of the back? How many are there altogether in the skin?

14. Describe the sweat. Why is it variable in amount? What is its average quantity? What is "insensible" perspiration? Sensible?

is poured out: then it is named *insensible perspiration.* When it is poured out faster than it evaporates, sweat collects in drops on the skin, and especially in regions, as the forehead, where the glands are very abundant. It is then named *sensible perspiration.*

15. **The Chief Use of the Sweat-Glands** is to cool the body when there is danger of its becoming too hot. Whenever a liquid dries up or evaporates, it draws heat from whatever it lies on. Your saliva is warmer than the skin of your finger, but if, after wetting your finger with this warmer liquid, you expose it to the air it feels cool, because as the moisture dries up, it takes heat from the skin. Our bodies only work well when they have a temperature of about 98 degrees of an ordinary or *Fahrenheit* thermometer. If they get a little hotter than this, *fever* results, and many organs either cease to work, or work very badly. In health, the sweat-glands pour additional liquid on the skin on a hot day, or when we are heated by exercise; and the heat taken away in the evaporation of this extra quantity of water, keeps the body cooled down to its proper temperature. When the body is cold the sweat-glands (except in disease) almost cease to make perspiration, and so loss of heat is checked.

16. **The Sebaceous or Oil Glands** of the skin, pour their secretion into the sides of the pits in which the roots of the hairs are contained. The openings of the ducts of a pair of sebaceous glands are seen at *k* in Fig. 22. The

15. What is the chief use of the sweat-glands? Explain. What is the proper temperature of the body? If it become a little hotter what results? When do the sweat-glands secrete freely? When little? Why?

16. Where do the sebaceous glands pour their secretion? What are its uses? How do the sebaceous glands sometimes cause black spots in the skin?

oily matter which they produce is partly spread over the
hairs, and partly over the general surface of the skin.
A healthy skin is always a little greasy, except on the
palms of the hands and the soles of the feet, where there
are no hairs and no oil-glands. ۰ This oily covering prevents the skin from absorbing water when the air is
damp; and from drying up on a hot dry day.

Sometimes the mouth of a sebaceous gland gets choked
up, and then its secretion collects in it and becomes dark-
colored.   In this way little black specks are formed on
the skin.   They are most frequent on the nose, where
the sebaceous glands are large, though the hairs to which
they belong are very small.   Pressure between the finger-
nails will usually easily force out the collected secretion
and remove the speck.

17. Summary —The skin forms a tough, elastic, protective covering for the body, and is also the main organ
of the sense of touch.

It is loosely attached to parts below it, so that it can
glide over them in our movements, without being
stretched or torn.   Beneath it is a cushion of fat which
protects the muscles from injury by blows.

Another use of the fat beneath the skin is to check
loss of heat from the body.   Warm-blooded animals
which live in the sea, as whales, have a very thick layer
of fat, known as "blubber," under the skin.   This
enables them to retain their animal heat in spite of the
cold water around them.   For the same reason, a fat person can bathe longer in cold water without being chilled
than one who is thin.

17. Functions of the skin ?   How attached to parts beneath ?   Uses
of the fat under the skin?   Illustrate.   Layers of the skin ?   Characters

The skin consists of two layers. The outer or *epidermis* is not sensitive and contains no blood. The *outer part of the epidermis* is dry, horny, and dead. It is constantly being shed.

The *deeper layer of the epidermis* is moist, and, being well nourished by liquid soaked up from the blood-vessels of the dermis, grows fast and makes new cells which are pushed on towards the surface to take the place of those of the outer layer which are shed or rubbed off.

The *dermis* contains many nerves and much blood. An important use of the epidermis is to cover and protect it. If there were no epidermis, our whole surface would feel like the "raw" skin at the bottom of a blister.

Nails are made by great growth of the horny layer of the epidermis. Their function is protective.

Hairs are scattered over nearly the whole skin. They are threads of epidermis developed from the bottom of little pits in the skin. When large and abundant, as on the head, they protect from cold and blows. Even when small and few, they are useful by aiding in the sense of touch.

Glands are hollow organs which make special liquids. The liquid made by a gland is called a *secretion*, and is poured out through a tube, the *duct* of the gland, on some surface outside or inside the body. The sweat-glands and the oil-glands pour their secretion on the skin.

The *sweat-glands* are most active when the body is warm, and help to keep it at its proper temperature.

of outer layer of epidermis? Deeper layer? Characters of dermis? How epidermis protects the dermis? Nails? Hairs? Uses? Glands? Secretions? Ducts? Glands of the skin? Use of sweat-glands? Of oil-glands?

The *oil-glands* pour their secretion into the hair-folli_cles. They provide a sort of natural hair-oil. Their secretion, also, becomes spread over the skin and makes the cuticle slightly greasy, so that water tends to run off instead of soaking into it.

CHAPTER VII.

HYGIENE OF THE SKIN.—ANIMAL HEAT.—CLOTHING.

· 1. Why the Skin should be Kept Clean.—A film tends to collect on the skin daily. This film consists chiefly of dry dead cells from the surface of the epidermis, of oily matter from the sebaceous secretion, and of dust and dirt. A certain amount of sebaceous secretion is useful for reasons already given; but it may collect in harmful excess. If the accumulation of the above-named matters is not regularly removed from the skin it tends to choke the mouths of the sweat-glands, the so-called "pores of the skin," and interfere with their activity. These glands not only serve to regulate the heat of the body, but separate waste matters from it, among them a considerable amount of water which has served its purpose and needs to be removed. If the sweat-glands do not work well, other organs, the lungs and kidneys, whose duty it also is to remove water and wastes, have too much work thrown upon them. The entire skin should therefore be washed every day, except that on the top of the head. The hair takes so long to dry that it is not usually prac-

1 Of what is the film composed which collects on the skin every day? Why should it be removed? What are the pores of the skin? The uses of the sweat-glands? What organs are overworked when the sweat-glands do not act properly? What other reasons are there for keeping the skin clean?

ticable to wash the head oftener than once or twice a week.

No doubt many folk go about in very good health with very little washing; contact with the clothes and other external objects prevents any great collection of dirt or dead epidermis on the surface of the skin. But apart from the duty of personal cleanliness imposed on every one as a member of society in daily intercourse with others, the mere fact that the healthy body can manage to get along under unfavorable conditions is no reason for exposing it to them. The evil consequences may be experienced any day, when something else throws another extra strain on the already overworked lungs and kidneys.

2. Bathing.—A bath not merely cleanses the skin but, when of the proper temperature and taken at the right time, strengthens and invigorates the whole body. For strong healthy persons, a cold bath is the best. When the water is very cold they may take the chill off, but should not make it warm. For the delicate, tepid baths may be preferable, but should be very brief. Immediately after a bath the skin should be dried and well rubbed.

3. Effects of a Cold Bath—The first effect of a cold bath is to drive blood from the skin and make it pale. This is soon followed by a reaction in which the skin becomes red and full of blood, and a pleasant glow of warmth is felt in it. The proper time to come out of a cold bath is during this reaction. If the stay in the water be too long, the skin again becomes pale and blood-

2. What useful purposes are served by bathing? Proper temperature of bath? What should be done immediately after bathing?

3. Why does a plunge in cold water make the skin pale? What follows? When should one come out of a cold bath? What are the

less, and the person probably feels chilly, depressed, and uncomfortable for some hours. The bath has then done harm; it has weakened instead of bracing the system.

How long one may remain in cold water with benefit, varies with the temperature of the water and with the vigor of the person. A strong man can set up a healthy reaction after a much longer stay in the water than can a feeble one. Also a person used to cold bathing may with safety continue his dip longer than one unaccus· tomed to it. Many persons who have been taking warm or tepid baths all the rest of the year, injure themselves in the summer by commencing at once to bathe for twenty minutes or more in the sea. Such persons com- plain that sea-bathing does not agree with them; if they would begin with three or four minutes in the water, the first day, and gradually increase the time, they would in most cases be benefited.

4. When to Bathe.—It is perfectly safe for a healthy per- son to take a cold bath when warm, provided the skin is not perspiring profusely. On the other hand, never take a cold bath when you are feeling chilly; or when you are much fatigued either in mind or body. Under such cir- cumstances, the proper reaction is apt not to occur. A cold bath should not be taken soon after a meal, for the blood is then wanted in the digestive organs (Chap. XI.) and cannot be spared to the skin to set up the after-glow. For a brief daily dip, there is no time so good as imme- diately after rising, while the body is still warm from bed and in a rested vigorous condition.

consequences of staying too long in it? State the conditions which determine the length of time it is wise to bathe in cold water.

4. When is it safe to take a cold bath? When unwise? What are the best times for bathing?

5. Shower-Baths take less heat from the body than other cold baths. The falling water also stimulates the skin and aids in producing the after-glow. Hence shower-baths are valuable to those not in very good health, *provided they suffer no unfavorable reaction.* But the sudden shock is unfavorable to many people; especially to those having any difficulty with the heart. Persons with whom shower baths agree, frequently find it advantageous to stand with the feet in tepid or warm water while taking them, and to keep the head covered by an oilskin cap. They thus avoid headaches and cold feet, while getting the general benefit of the bath.

6. Warm Baths cleanse the skin more readily than cold, and are desirable once or twice a week for this purpose. Daily warm baths should not be taken except on medical advice. While promoting the tendency to perspiration, which is often important in the treatment of disease, they also, when frequent, diminish the general vigor of the body.

7. The Use and Abuse of Soap.—Nearly all soaps contain so much potash or soda that lathers made from them are really weak "lye." On this their main cleansing power depends; for, like the lye used to remove stains from floors, they take up greasy matters and make them capable of being washed away by water. The potash or soda of soap often does harm, causing too free removal of the oily sebaceous secretion, a thin layer of which is necessary to protect the skin from too rapid

5. In what circumstances are shower-baths desirable? What precautions should delicate persons observe in taking them?
6. What are the uses of warm baths? Their dangers?
7. What makes soap cleansing? How may it injure the skin? How should its use be limited? Name a good substitute for soap.

drying. Probably as many skin-diseases have been caused by too free use of soap, as by uncleanliness. Except on parts of the body especially exposed to contamination, soap should not be applied oftener than twice a week. More frequent employment of it is quite unnecessary for cleanliness, if a daily bath, followed by a good rubbing with the towel, be taken. Persons whose skin is injured by even the occasional use of soap, will find in corn-meal a good substitute.

8. Cosmetics and Hair-Dyes.—When the face is hot and perspiring, a good deal of comfort may often be obtained by applying a little finely powdered arrowroot, and immediately wiping it off with a dry towel. This is better than plunging the face in water, which often causes it to become more flushed afterwards. No face-wash, whitening, rouge, or other coloring matter should ever be used. In spite of the assertions of their makers, which induce foolish folks to buy them, nearly every one contains materials highly injurious to the skin. Those which do not, are hurtful by interfering with the proper growth of the epidermis and by checking the action of the sweat-glands. Many face-washes contain poisons which, being absorbed by the skin, ruin the health.

Most hair-dyes contain lead or some other poison. As they are kept off the skin as much as possible, they do not in most cases injure it, but they always harm the hair, never improve its appearance, and seldom succeed in their purpose of deceit.

8 State a harmless method of quickly cooling a heated face. Why should face-washes and other "cosmetics" be avoided? Why hair-dyes?

9. Burns and Scalds.—If slight, cloths soaked in strong solution of bicarbonate of soda (common washing or cooking soda) may be applied, and renewed when they begin to dry. This greatly relieves the pain. If the burn or scald be deep and extensive, endeavor to exclude the air and prevent rubbing until medical aid can be obtained. The best application for these purposes, is raw cotton soaked in a mixture of linseed-oil and lime-water in equal parts. If this is not at hand (as it should be in every house distant from a drugstore), sweet-oil or fresh lard may be used instead.

10. Action of Alcoholic Drinks on the Skin.—Taken into the body in any form, alcohol causes more blood to flow to the skin. This is seen in the flushed face of a man who has been "drinking." If the drinking be continued, the redness becomes permanent. The skin is then puffy and congested, and the face especially acquires a reddish blotchy look. Its proper nourishment being interfered with, the epidermis collects in scaly masses. The peculiar degraded look of the sot's face is the result.

11. Animal Heat.—Sometimes you feel hot, sometimes cold. This feeling is due to changes in your skin. The mouth may feel hot after drinking a cup of tea, or cold for a short time after eating ice-cream: but this does not make us say that *we*, that is our bodies in general, feel warm or cold.

9. What handy remedy is useful for slight burns? What should be done in case of severe burns or scalds?
10. How does "drinking" affect the skin? The expression?
11. To what are "feeling hot" and "feeling cold" due? Illustrate. What is "animal heat"?

If you keep your mouth closed, your tongue does not feel warm on a hot day, or cold when the air around you is at a freezing temperature. The reason of this is that in your body heat is being produced all the time, keeping the internal parts warm. This heat is known as *animal heat.*

12. The Temperature of the Body.—So long as you are in health, a thermometer placed in your mouth would indicate almost exactly the same temperature every day in the year. This is a very curious fact. A stone or a frog is cold on a cold day and warm on a hot day; but, except sometimes on the outside, your body is always hot, and hot to very nearly the same degree; in health never below 98° or above 101° of an ordinary Fahrenheit thermometer. All animals, as birds and beasts, which, like man, have a regular temperature of their own, are known as " warm-blooded " animals. Any condition of the body in which its organs are hotter than their proper temperature, is known as a "fever."

13. How the Body is kept from getting too Hot.—Everything that works, even two sticks rubbed across each other, produces heat, though in many cases it is too slight to be noticed. The organs of our bodies are no exception; and the more they work, the more heat they produce. If all this heat remained in the body, we should soon be in a high fever. It is carried off in several ways.

12 What does a thermometer placed in the mouth show? How does your body differ as to temperature from a stone or a frog? What is the healthy temperature of the interior of the human body? What is meant by "warm-blooded animals" ? What is fever?

13 Where is heat produced in the body? When most? Why must some of it be got rid of? How do the lungs help in keeping us from becoming too warm? The sweat-glands? The blood flowing through the skin?

Partly, for example, by the air we breathe out, which is nearly always hotter than the air we breathe in: and so carries heat away from the body. But the skin does more than any other organ in regulating the bodily temperature.

The skin gets rid of the heat in two ways. In the first place, its glands produce perspiration, and the evaporation of this perspiration, as we have already learned, carries off heat (p. 67). We thus see why it is useful that we perspire more freely on a hot day, or when we are exercising and our muscles producing a great deal of heat.

In the second place, except on the very hottest summer days, the air around us is cooler than the inside of our bodies. Blood which has been made hot as it flowed through the internal organs, is sent to the skin and there, giving heat to the air, is cooled. It is then carried back from the skin to the inside, picks up more heat, flows again to the surface and gets rid of it; and so on, all the time.

14. How the Body is kept from getting too Cold.—The fat which lies beneath the skin may be compared to the packing in the sides of a refrigerator. It checks the passage of external heat or cold to the inside. Accordingly, thin persons cannot bear exposure to cold as well as those who are fat. Too great loss of heat is also prevented by the diminished activity of the sweat-glands in cold weather, and by the fact that most of the blood is then kept away from the skin, which accordingly becomes pale. An exception to this is found when there has been

14. How does fat aid in keeping us from too great cooling ? How do the sweat-glands behave in cold weather? Why does the skin usually become pale in a cold room? When may the skin be red and perspiring even in cold weather ?

great production of heat in the internal organs. Then the sweat-glands act, and the skin becomes full of blood even on a winter's day. You know that if you sit still in a cold room your skin becomes pale and you do not perspire; while, so long as you are in health, a good run in the coldest weather will flush the skin and cause perspiration.

15. Clothing.—Clothes are employed by mankind for many purposes of ornament and ostentation; and these unimportant uses are sometimes allowed to interfere very seriously with their main objects. The real uses of clothing are physiological and hygienic. These uses are, (1) to aid the skin in regulating the temperature of the body; (2) to protect the skin itself from too rapid heating or cooling; (3) to prevent a sudden rush of blood from the skin to internal organs when the air around the body is quickly cooled.

16. What Properties the Materials used for Clothing should Possess.—Nature has provided all warm-blooded animals who thrive in parts of the earth where the climate is variable, except man, with a covering of fur or feathers. This covering becomes thicker in the cold seasons of the year, and thinner in the warm. It also is made of materials which greatly hinder the passage of heat through them. Fur and feathers are accordingly known as *bad conductors* of heat. In winter they check

---

15. For what unimportant uses is clothing employed? What its real uses?

16. What clothing does Nature provide for most warm blooded animals? How does it change with the seasons? What are the properties of its materials as regards the transmission of heat? How do fur and feathers protect the skin from sudden changes of temperature? What lessons should man learn from the clothing provided for lower animals by Nature?

loss of heat from the skin; and all the year round they keep the skin from being rapidly cooled or heated when exposed to sudden changes in temperature.

Man has to provide his own clothing, but should always bear in mind this lesson from Nature: His clothing should vary in amount with the season, but the chief garments should be made of materials which are bad heat-conductors.

17. The Relative Value of various Clothing Materials.— *Furs* are very bad conductors, and do not easily become damp. They are the most suitable clothing for very cold weather. *Woollen fabrics* are also excellent. *Silk* comes after wool, and in our variable climate forms the best material for the underclothing of those whose skins are irritated by woollen materials, such as merino. *Cotton* is not so good as silk, but is far better than *linen*, which not only allows heat or cold to pass readily through it, but easily absorbs moisture and becomes damp. The same objection holds against linen bed-clothing. Cotton should be used, except, perhaps, for pillow-cases in summer.

The proper clothing will vary with climate and season; but, except for those living in regions where sudden temperature-changes do not occur, the following is the proper rule: Wear silk or wool next the skin; over this regulate the amount of clothing according to the weather, but always wear at least one other covering of non-conducting material, cloth, silk, or flannel.

17. Name common materials for clothing in order, putting the worst conductors of heat first. Why is linen not so good a clothing material as cotton? What rule as to clothing should be observed by all who live in a variable climate?

CHAPTER VIII.

FOODS.

1. How the Body is Built up and Repaired.—So long as you are growing, you require a supply of material out of which your body can make more bone and muscle and skin and blood, and the rest. This material is supplied in the things you eat and drink.

Even after a man is full-grown, he still needs a quantity of food daily, to repair his body. Every time an organ works, some of it is used up and turned into useless waste things, which are soon carried away from the body through the pores and other outlets. If they are kept in it, as they sometimes are in disease, they clog all the organs and interfere with their work. If a man be starved, he becomes lighter every day, because he makes waste matters, and these are carried away by the skin or the lungs (Chap. XV.) or the kidneys (Chap. XVII.) or other organs, all the time, so long as he lives.

2. The First Use of Foods is, then, to furnish materials for the building and repair of the body. In early life, the building exceeds the waste, and *growth* takes place.

---

1. What does the body require while growing? How supplied? Why does a full-grown man need food? Why must the wastes of the body be removed? Why does a man become lighter if he takes no food?

2. What is the first use of foods? Why do we grow while young? Why not in middle age? What often happens in old age?

Then comes a period of middle life, in which they are
about equal. Finally, in old age, it often happens that
the organs cannot make use of enough food for their
complete repair, and therefore slowly diminish in size.
The muscles and bone of an old man often become
"wasted" and feeble.

3. A Second Use of Foods is to give us strength and
keep up our animal heat. A starving man not only becomes lighter every day, but weaker and colder. This
use of foods may be compared to the use of coal in the
furnace of a steam-engine. And just as the coal would
be useless if it did not burn, and will not burn unless
there be a draught of air in the furnace, so foods would
neither make us strong nor warm did they not contain
things which could very gently burn inside the body;
and in order to burn, these things must be supplied with
air by our breathing.

4. Oxidation.—The air which we breathe into our
bodies is a mixture of two gases; only one of them is
useful in keeping a fire alight or in burning foods
inside our bodies. It is named *oxygen.* Generally
when anything burns, it unites with oxygen The
thing burned is then said to be *oxidized*, and the process
of uniting with oxygen is named *oxidation.* When oxidation takes place very fast, a great deal of heat is given
out along with light, as in a fire or candle. But oxidation sometimes goes on slowly; and then the tempera-

3. What is the second use of foods ? How shown ? To what may
the use of foods be compared ? In order that a fire may burn what is
necessary besides coal or wood ? What must food contain ? What
is necessary that these things may burn ? How is it supplied ?
4. What is oxygen ? When is anything said to be oxidized ? What
is oxidation ? What are the effects of rapid oxidation ? Of slow ?
Which kind of oxidations occurs in our bodies ? Why ?

ture does not become very high and no light is produced. The oxidations which take place in our bodies, are of course slow oxidations; otherwise they would burn us to ashes.

5. Examples of Slow Oxidation.—A good example of a slow oxidation is afforded by the rusting of iron; this is an oxidation; and the rust is iron united with oxygen. This oxidation usually occurs even more slowly than those which take place in our bodies, and heat is given off so slowly that rusting iron does not feel warm when we touch it. You know, too, that iron rusts easily in damp air, and in this respect the oxidation of the iron is like the oxidations which occur inside our bodies, which are moist in every part.

6. Definition of Foods.—*Foods include all substances, except air, taken into the body and serving for any one of three purposes—*(1) *to provide material for its growth or repair, or* (2), *by their oxidation, to supply it with working power or to keep it warm, or* (3) *to aid in carrying nourishment from part to part.* To the above, we must add that for a substance to be properly a food, *neither itself nor anything produced from it inside the body shall be injurious to the structure or action of any organ;* otherwise it would be a *poison,* not a food.

Most foods serve more than one purpose Thus meat and bread furnish material for growth and repair, and also supply strength and warmth. Water is found in all the organs, and is a necessary part of them; but it also

---

5. Give an example of slow oxidation. In what other respect does the rusting of iron resemble the oxidations which take place in our bodies ?

6. Define foods. Poisons. What purposes are served by meat and bread ? By water ? Illustrate.

dissolves solid foods and carries them into the blood to be conveyed to places where they are needed. A lump of sugar in your mouth would not nourish you, unless the saliva should dissolve it, and then it should be taken up into the blood and borne to muscle or brain or skin, or some other part that wanted new material.

7. Classification of Foods.—Foods, like the body itself, consist of both things which will not burn or oxidize, and things which will. The food-materials which will not burn, such as water and common salt, are known as *inorganic foods*. The foods which will burn, are obtained either from animals or plants, and are named *organic foods*, because they are obtained from living things, which have *organs*.

8. Most Foods contain more than one Nourishing Substance.—Beef, for example, contains (1) water, which goes off when we dry it; (2) certain minerals, which are left in the ashes if the meat be burned, and which, when meat is eaten, are useful in building bone, blood, muscle, and brain; (3) organic matters of several kinds: the fat is one of them; another, found in the lean, is named an *albumen*. It is in nearly all respects like the white of an egg. To take another example, wheaten bread contains (1) some water; (2) minerals; (3) a kind of albumen; (4) starch; (5) a little fatty matter. Each nourishing substance found in any food, is named a *food-stuff*.

9. The Chief Kinds of Organic Food-Stuffs are—(1) albu-

7. Of what do foods consist? What are inorganic foods ? Examples. Organic? From what obtained ? Why so named ?

8. What substances, useful to the body, does meat contain ? What does the albumen resemble? What are the useful substances in bread ? What is a food stuff?

9. Name the chief organic food stuffs.

mens; (2) jelly-forming substances; (3) fatty or oily matters; (4) sugars; (5) starch. The albumens are the most important. A man can maintain life on water and lean meat, while if he should get, along with plenty of water, all the fat and sugar and arrowroot (which is nearly pure starch) that he could eat, he would slowly starve.

The reason of this is very simple. A special substance, named *nitrogen*, is essential for the making or repairing of all the organs of the body. Albumens contain some of this substance; fat, starch, and sugar do not. Every day some nitrogen is carried away from the body in its waste matters. If none of it be supplied in the food, a man will therefore slowly die of nitrogen-starvation, no matter what abundance he may have of other things.

Crackers and cheese would be useless to a man dying of thirst; so fat or sugar or starch would be useless to a man whose organs were starving for nitrogen.

10. Inorganic Food-Stuffs.—A sufficient quantity of most of these is contained in bread and meat and milk and our other common foods. Thus iron is an essential part of the blood, but in health we need no more than is contained in the vegetables and meat which we eat daily.

Water and common salt are the only inorganic food-stuffs that are usually taken by themselves or specially added to our food in cooking. The body daily gives off more of each than it would otherwise receive.

Which are the most important? How do we know that the others are less valuable? Explain why starch and fat cannot take the place of albumens in nourishing the body. Illustrate.

10. How are we supplied with most inorganic food-stuffs in sufficient quantity? Illustrate. How are water and common salt exceptional? Why?

11. Common Salt is found in every solid part and every liquid of the body. It has been maintained that salt as a separate article of diet is a mere luxury, and there seems to be some evidence that certain savage tribes live without more than they get in the meat and vegetables which they eat. There is, however, no doubt that.to many animals, as well as most men, the want of salt is a terrible deprivation. Buffaloes and other creatures are well known to travel miles to reach " salt-licks;" of two sets of oxen, one allowed free access to salt, and the other given none save what existed in their ordinary food, it was found after a few weeks that the former were in much better condition. In man the desire for salt is so great that in regions where it is scarce, it is used as money. In some parts of Africa, a small quantity of salt will buy a slave, and to say that a man commonly uses salt at his meals, is equivalent to stating that he is a luxurious millionaire.

12. Meats, whether derived from bird, beast, or fish, are highly valuable foods. They supply material for making tissues, for providing working power, and for keeping up animal heat.

13. Milk will support life longer than any other single food. It contains water, minerals, a kind of albumen named *casein*, which, when separated, forms cheese; fatty matters, especially *butter;* and a sugar named *milk-sugar*. In milk there is more lime than any other common food; it is therefore very valuable in childhood when the bones are growing rapidly.

11. In what parts of the body is common salt found ? What is said to result from want of it ? Illustrate the natural longing for salt ?
12. What is said of the value of meats ?
13 Why is milk a very valuable food ? Name the chief food-stuffs contained in it Why is it especially valuable in childhood ?

14. Eggs are rich in albumen and fats. They contain a great deal of valuable nourishment in a small bulk. The white, or albumen, is more easily digested when cooked soft, and the yolk when cooked hard. So in the old controversy about hard-boiled and soft-boiled eggs, as in a good many controversies, both sides are wrong and both sides are right.

15. Bread made from wheaten flour is more nourishing than any other, as it contains, besides much starch and a little sugar and fat, a good deal of a kind of albumen named *gluten*. In preparing ordinary white flour, the husk of each grain of wheat is sifted out by a process known as *bolting*. This husk contains a good deal of nourishing matter. In unbolted flour this is saved. Many persons also find bread made from it more wholesome than that made of bolted flour. In other cases it unduly irritates the bowels. *Maize* or *corn* contains more starch and fats than wheat, but much less albumen.

16. Vegetables and Fruits.—*Rice* contains a great deal of starch but hardly any albumen: by itself it is a very poor food, but taken with food rich in albumen, as meat of any kind, it is excellent. *Peas* and *beans* are good foods: they contain much albumen and starch. *Potatoes* are a poor food. Other fresh vegetables, as *cabbage, turnips*, and *tomatoes*, are useful mainly for the mineral matters contained in them. Most of their weight is due

14 What do eggs contain? When are they more easily digested?
15. What is the most nourishing kind of bread? Why? What is meant by the "bolting" of flour? What is saved when flour is unbolted? What is said of the healthfulness of eating bread made from unbolted flour? How does corn differ in composition from wheat?
16. What is said of rice? Of peas and beans? Potatoes? Other fresh vegetables? What is their chief constituent? What is said of fruits? Give proof of their value.

simply to water; organic food-stuffs are present in them in very small quantity. *Fruits*, like most fresh vegetables, are chiefly valuable for their mineral matters. Some kind of fruit or vegetable is, nevertheless, an important part of every one's diet. This is shown by the fact that sailors on a long voyage almost invariably suffered from the disease known as *scurvy*, before the "canning" of vegetables and fruits made it possible to keep the crew supplied with them.

17. Jelly.—Jellies made from animal substances, as calves' feet, or the *gelatin* sold in groceries, are commonly believed to be extremely nutritious. It is therefore important to know that, although they contain nitrogen, they cannot entirely take the place of albumens. When our bodies are supplied with animal jelly, they can manage to get along with less albuminous food, but the organs need for their growth or complete repair, food containing some albumen. If a sick person can digest some lean beefsteak, it is more valuable as a food than the best calf's-foot jelly; but if he can only digest the jelly it is very useful, because, though it does not entirely prevent the loss of nitrogen from the body, it considerably lessens it.

18. The Cooking of Meats, in many cases causes special flavoring matters to be formed, which make our food more palatable. In addition it makes many foods more digestible.

When meats are properly cooked they become softer

17. What jellies are usually supposed to be very nourishing ? What is it important to know about them ? How and when may they be very useful?
18. How does cooking make meats more pleasant to the taste ?

and more easily broken up by the teeth because their connective tissue loses its toughness, being for the most part turned into jelly. If the meat be cooked too fast this change occurs very imperfectly, and it comes to table stringy, tough, unpleasant to eat, and hard to digest.

When meat is boiled, much of its flavoring and some of its nourishing matters are apt to pass out into the water and be lost. If the meat be plunged *at first* into boiling water for a few minutes, the surface is hardened and a coating formed, which keeps in the flavoring matters of the deeper parts. The cooking should then be continued slowly. Quick boiling, except at the start, will spoil the best and most tender piece of meat.

Hogs are especially apt to suffer from a parasite, which lives in their muscles. This parasite is a little worm named *trichina*. If the meat be eaten raw or imperfectly cooked, these parasites bore their way out of the alimentary canal and travel all over the body, producing the disease known as *trichinosis*. This danger may be avoided by thorough cooking, which kills the trichinæ.

19. The Cooking of Many Vegetables is very important. Those which are not eaten in the green state for their minerals, nearly all contain starch as their chief constituent. This starch exists in the form of tiny solid particles which are very hard to digest. When the vegetable is boiled, these particles are softened and made easier to

More digestible ? Why should they be cooked slowly ? Why should a joint which is to be boiled be put at first in very hot water? Why should the cooking be finished slowly? Why should hog-meat, especially, be thoroughly cooked ?

19 What is the chief nutritive substance in vegetables ? How is it altered by boiling ? By roasting ?

digest. When starch is roasted, it is converted into a substance known as *soluble starch* which readily dissolves in the mouth or stomach. The common belief that the crust of a loaf is more easily digested than the crumb, and toast than ordinary bread. is therefore correct.

CHAPTER IX.

1. What is Meant by a Stimulant.—In general a stimulant is something that does not nourish the body, but stirs it or one or more of its organs to do work. Thus we say that a man is stimulated to labor by the desire to make his family comfortable; or a lad to hard study by the wish to get to the head of his class; or to the use of his muscles to their utmost power, by the ambition to win a race. Some stimulus to exertion is useful: without it most of us would be slothful and ignorant and stupid. On the other hand, our bodies may be stimulated to attempt more than they can safely accomplish. Many a man breaks down from too severe labor, and boys and girls at school sometimes injure their health by overstudy, stimulated by the ambition to excel.

2. Foods as Stimulants.—Several common articles of diet are named *stimulants*, because their action is rather to excite the brain, or the heart, or the muscles, or the stomach to greater activity, for a time, or to decrease the feeling of fatigue after labor, than to nourish any organ. Some of these stimulants, as pepper, which makes many

1. What is meant by a stimulant? Illustrate. Why is some stimulus useful? How may it be an evil?

2 Why are several common articles of diet called stimulants? What is said of their different effects? Of those who need not even the least injurious?

foods more palatable, do little or no harm as ordinarily used. Others, as alcohol in all its forms, when taken at all are very apt not to be used in moderation, and then they do so much injury that they are really poisons. Persons in perfect health need no kind of stimulant food. A strong, healthy young person with rich blood, powerful heart, vigorous muscles, and good digestion wants no pepper nor mustard nor tea nor coffee to promote his appetite or relieve his fatigue. He is better without such things; and so is a perfectly healthy man or woman.

3. The Use of Stimulants.—Stimulant articles of diet are rather medicines than foods; as medicines they have their use. A man sometimes comes home after his day's work, fagged out in body and mind, without appetite, and feeling restless and jaded. Then a cup of tea will often remove the feeling of fatigue, enable him to eat and digest his supper, soothe his nerves, and let him get a good night's rest. The tea has not itself nourished him, but it has enabled him to take proper nourishment, and in that way has done good.

We may compare the safer kinds of stimulants, as tea and coffee, to the "blower" of a grate. When a fire is burning badly the blower is useful, but if the fire is burning well it only does harm. It leads to a very rapid using up of the coal or wood, without any corresponding benefit, and does not itself supply fresh fuel.

4. The Abuse of Stimulants is chiefly due to the fact that the brief relief from fatigue, and the feeling pro-

3. Rightly considered what are these stimulants? What is said of their effects when properly used? To what may they be compared?
4. To what is the abuse of stimulants chiefly due? What are the wrong and right ways of regarding them?

duced by them of being able to do more work, is taken as a sign that they have really strengthened the body. They come to be regarded as *foods* which may be taken safely so long as there is an appetite for them, and not as *medicines* to be taken always with caution.

5. Tea and Coffee.—The amount of nourishment contained in a cup of tea or coffee, apart from the sugar or milk put into it, is trivial. Both liquids have, however, a great power of making the brain tranquil, and of removing the feeling of fatigue or worry. When taken in moderate quantity, they rarely leave injurious after-effects. Some persons, however, experience a sensation of fulness in the head after taking coffee, or are kept awake all night by a small cup of it; they should of course avoid it. For relieving muscular fatigue, tea or coffee is far superior to any kind of alcoholic drink. Sportsmen out for a day's shooting find a flask of cold tea in the pocket far more useful than a flask of spirits. Generals who have commanded troops in campaigns agree that a ration of coffee is better than one of whiskey for tired soldiers. All commanders of arctic exploring expeditions have come to the conclusion that the men bear fatigue, cold, and anxiety better on tea or coffee than when supplied with rum or whiskey instead.

6. The Harm done by Excessive Tea- or Coffee-Drinking.—Injurious effects of excessive tea- or coffee-drinking are most commonly seen in those who are young, or who, though older, lead indolent lives. The conse-

5. What is the chief nourishment in a cup of tea or coffee? Effect of tea or coffee on the body? When should coffee be avoided? Effect of tea or coffee on muscular fatigue? Illustrations.
6 What class of persons are most liable to be injured by tea- and

quences of *excessive tea-drinking* are, dryness of the mouth, loss of appetite, biliousness, a feeling of sickness at the stomach, nervousness and unreasonable trembling, troubled sleep and terrifying dreams. In their full development, these symptoms are often met with in professional "tea-tasters;" but they are not unfrequent in idle men and women, who take no part or interest in the world's work and who strive to keep themselves from utter stagnation by drinking strong tea, morning, noon, and night.

Coffee taken in excess tends rather more than tea to dilate the channels through which blood passes to the brain. It then causes a feeling of "fulness" in the head and flushes the face. It is more apt to produce wakefulness than is tea; but its action on the digestive organs when it is taken in excess is not so bad. Some people have their digestion disturbed by coffee *with milk*, so that it gives them hazy vision, dizziness, and headache, while the same persons experience no harm from the same amount of coffee without milk.

7. Alcoholic Stimulants.—Young persons do not generally know what alcoholic stimulants are, and often suppose them to be only the various forms of "strong spirits," such as brandy, rum, gin, and whiskey. But all wines contain alcohol, and so do all beers, cordials, and even cider, except when it is perfectly new. Many of the "*tonics*" so widely advertised, are also alcoholic drinks, sold under a false name. We have already

coffee-drinking? Consequences of excessive tea-drinking? In whom most often fully seen? What other class of people are apt to exhibit them? How do the effects of excessive coffee-drinking differ from those produced by tea?

7. Name some drinks containing alcohol. Why is it obvious that

leained that alcohol tends to injure seriously the connective tissues, the muscles, and the skin. We shall later learn that it acts quite as injuriously on many other parts of the body; for example, the heart and the brain and the lungs. It is thus obvious that all drinks containing alcohol are dangerous, and the more so the greater the quantity of alcohol in them. For the present, we will confine ourselves to the question whether alcohol has any just claim to be called a food. Foods are useful to build tissues, to supply strength or working power, or to maintain our animal heat. Does alcohol do any one of these?

8. Is Alcohol a Tissue-Forming Food?—To this the answer is certainly, *no;* so far at least as useful tissue is concerned. Its consumption often leads to excessive and harmful overgrowth of connective tissue and fat, but it does not lead to development of muscle or brain or gland.

9. Is Alcohol a Strengthening Food?—To this the answer is also *no.* Alcohol in small doses is a stimulant to brain and muscle, and may for a short time excite them to overwork or to work when they should be resting. But as it nourishes neither of them, the final result is bad. The brain and muscle are left in an injured state. As regards the brain, the consequence is often insanity (Chap. XIX.). As regards the muscles, very careful experiments have been made on soldiers who

alcoholic drinks are dangerous? In deciding the claims of alcohol to be a food, what properties of foods must we recall?

8. What is said of alcohol as a tissue-forming food?

9. Is alcohol a strengthening food? How may it lead to overwork? Results? What were the results of experiments made on soldiers as to the action of alcohol on the muscles?

were given definite tasks to accomplish. The result was that on the days on which they were supplied with spirits, they could neither use their muscles as powerfully, nor for as long a time, as on the days when they got no alcoholic drink.

10. Does Alcohol keep up the Heat of the Body?—To this question, also, the answer is *no*, though this may seem strange in view of the fact that a drink is often taken " to warm one up." The apparent inconsistency is easily explained. We have already learned that our feeling of being warm depends on the nerves of the skin (p. 76). We have no nerves which tell us whether heart or muscles or brain are warmer or cooler. These inside parts are always hotter than the skin, and if blood which has been made hot in them, flows in large quantity to the skin, we feel warmer because the skin is heated. As alcoholic drinks make more blood flow through the skin, they often make a man feel warmer. But their actual effect upon the temperature of the whole body is to decrease it. The more blood that flows through the skin, the more heat is given off from the body to the air, and the more blood so cooled is sent back to the internal organs. The consequence is that alcohol cools the body as a whole, though it may for a short time heat the skin. That a large dose of alcohol leads to excessive loss of heat from the body, has been thoroughly proved by many observations on drunken men, and by experiments on the lower animals.

10. Does alcohol maintain the heat of the body? Why does a drink sometimes make a person feel warmer? What is the real effect of alcoholic drinks on the temperature of the body? How has it been proved?

11. Alcohol is a Medicine.—In many diseases the body needs rousing to make a special effort, and the physician has to order an alcoholic stimulant, or some substance which acts on the body in a like way. Not merely is it a powerful stimulant, but in moderate doses it checks for a time the oxidations of the body, and thus diminishes the wasting of its organs. This only harms a healthy well-nourished person, but may often be useful in impaired health. In fact, alcohol is a medicine, and often a very valuable one. It may be classed with strychnine, arsenic, opium, and other drugs, which are useful in various impaired states of health, but so dangerous that they should only be taken on the advice of a doctor, and in the exact manner and quantity ordered by him.

Probably few physicians would be willing to omit alcohol from the list of remedies; but many patients have acquired drinking habits from first taking an alcoholic stimulant on the "doctor's advice." Many medical men, for this reason, prescribe it in some disguised form; and this is the better plan.

11. When may it become necessary to take alcohol internally? With what poisonous drugs classed? Precautions to be used?

CHAPTER X.

DIGESTION.

1. Introductory.—We learned almost at the outset of our anatomical study, that the alimentary canal is but a tube (Fig. 1) which, beginning at the mouth, runs through the neck, chest, and abdomen, and ends by opening again on the outside at the lower part of the trunk of the body. We now have further to observe that it is wide in some parts, like the stomach and large intestine (all to be presently described), and narrow in others, like the gullet and small intestine, which will also be presently described; some parts of it are straight and others coiled; but it has no branches which reach out into the arms or legs or brain. Nevertheless, after a good dinner we feel no doubt that what we have eaten is going to strengthen our limbs and every other part of the body. To accomplish this, the nourishing portions of the food must get through the walls of the alimentary canal, and then be carried to all the organs.

2. Digestion—The first important thing that happens to our food inside the alimentary canal, by way of preparing it to reach distant organs, is that its solid parts, or at least those of them which are nourishing, are dis-

1 How do the various parts of the alimentary canal differ? What must happen in order that food may nourish all parts of the body ?
2. What is the first important work inside the alimentary canal to

solved. This is brought about by the action of peculiar liquids made inside the body, and poured into the mouth, stomach, or intestines. The process of getting all the valuable part of the things which we have eaten, into a liquid state, is known as *digestion.*

, **3. Absorption.**—The second step is to get this nourishing liquid into the blood. As it slowly passes along the alimentary tube it is gradually soaked up or *absorbed* by the walls of the latter, as if they were lined with blotting-paper, and either mixed at once with the blood which flows in them; or, first, with another liquid, the *lymph*, which is afterwards poured into the blood. The taking up of digested food by the lining of the alimentary canal, is known as *absorption.*

4. The Alimentary Canal is about thirty feet in length, much the longest portion of it being contained in the abdomen. At its beginning (Fig. 24) it is tolerably wide and forms the *mouth* and *throat cavities.* In the neck and chest, it has the form of a narrow, nearly straight tube, the *gullet* or *œsophagus.* The lower end of the gullet passes through the diaphragm and then almost immediately opens into the much wider *stomach* (Fig. 32). The stomach is followed by the narrow greatly coiled *small intestine;* and this in turn opens into the *large intestine,* which is the last portion of the alimentary canal.

prepare food for nourishing the body ? How brought about ? What is the process called ?

3. What is the second step in digestion ? How does the nourishing liquid get into the blood? What is lymph ? What is this process of * taking up digested food called ?

4. How long is the alimentary canal? What easily contains the longest portion of it? What is the gullet ? Where does it end? What follows the stomach ? What is the large intestine ?

5. The Lining of the Alimentary Canal is a soft, red, moist kind of skin, named a *mucous membrane.* You can easily see part of it on examining the inside of your mouth with the help of a looking-glass. This mucous membrane has two functions, *secretion* and *absorption.* Imbedded in it are thousands of tiny *glands* (p. 65), which, instead of making perspiration or oily matter, like the skin-glands, pour out very different liquids, which aid in swallowing and digesting.

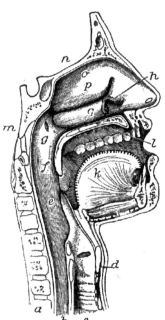

Fig. 24. — The mouth, nose, and pharynx, with the commencement of the gullet and larynx, as exposed by a section, a little to the left of the middle of the head. *a,* vertebral column; *b,* gullet; *c,* windpipe; *e,* epiglottis; *f,* soft palate; *g,* opening of Eustachian tube; *k,* tongue; *l,* hard palate; *m,* the sphenoid bone on the base of the skull; *n,* the fore part of the skull-cavity; *o, p, q,* the turbinate bones of the outer side of the left nostril-chamber.

6. The ✻ Mouth-Chamber (Fig. 24) opens in front between the lips, and behind into the *throat-chamber* or *pharynx.* It is bounded on the sides by the cheeks, below by the tongue, above by the *palate.* The front portion of the palate, *l,* separates the mouth from the nose, and is supported by bone. This portion is named the *hard palate.* The posterior portion of the palate, *f,* is soft

5. What is the lining of the alimentary canal? Where can you easily see it? What are its functions? How does it aid in swallowing and digesting?

6. Describe the mouth-chamber. The hard palate. The soft pal-

and contains no bone. It forms a curtain between the
mouth and pharynx; there hangs down from its lower
border a soft fleshy projection, named the *uvula,* generally miscalled the palate. If the mouth be held wide
open in front of a mirror, the uvula can be easily seen,
and also the opening, between the soft palate and the
root of the tongue, which leads into the pharynx. This
opening is named the *isthmus of the fauces.* On its sides
are the *tonsils.*

7. **The Teeth** stand almost alone among the organs
of the body, in the fact that when broken or seriously
injured or much worn, they are not repaired. To
do their duty they must be very hard, and they gain
this hardness by being so largely made of mineral
matter that their living animal part is not present in
sufficient quantity to rebuild them when they are broken
or decayed. During life two sets of teeth grow. The
first, named the *milk-teeth,* are developed and shed during childhood. The second set, named the *permanent
teeth,* follow the milk-teeth. If they are lost, we must
go to the dentist, for no others will grow to take their
places.

8. **The Forms and Uses of Different Teeth.**—Every tooth
consists of a *crown,* the part which projects into the
mouth; of a narrower *neck,* surrounded by the gums;
and of one or more *roots* or *fangs,* tightly fitted into pits
(called *sockets*) in the edges of the upper and lower jaw-

ate. What is the opening seen between the soft palate and root of
the tongue? The organs on each side?
7. How do the teeth differ from most other organs as to repair?
How is this accounted for? What is said of the first set of teeth?
The second?
8 Of what parts does a tooth consist? Give names of the different
teeth. Describe the incisors, Canines. Molars. Bicuspids.

bones. On account of differences in the shape of their crowns, and in their uses, the teeth are divided into *incisors, canines, bicuspids,* and *molars.* The incisors (Fig. 25) have sharp chisel-shaped edges and are adapted for cutting our food. The canines (Fig. 26) or *eye-teeth* are

Fig. 25.   Fig. 26.   Fig. 27.   Fig. 28.

Fig. 25.—An incisor tooth.
Fig. 26.—A canine or eye tooth.
Fig. 27.—A bicuspid tooth seen from its outer side; the inner cusp is accordingly not visible.
Fig. 28.—A molar tooth.

pointed and serve the same purpose: they are very long and sharp in dogs and cats, and are useful to these animals in holding their prey. The molars (Fig. 28) have broad rough ends to their crowns and are suited to grind and crush. The bicuspids (Fig. 27) are like the molars but not so large.

9. Arrangement of the Teeth in the Jaws.—In the milk-set, there are twenty teeth, ten in each jaw. Beginning in the middle line and going back, we find in order, on each side, two incisors, one canine, two molars.

The permanent teeth number sixteen in each jaw. Beginning at the middle line, we find successively two incisors, one canine, two bicuspids, and three molars, in each half of each jaw. The incisors and canines take the places of the milk-teeth of the same names. The

9. Arrangement of milk-teeth. Of permanent teeth. Which ones are added as the jaw grows larger? What of the wisdom-teeth?

bicuspids supplant the milk-molars. The permanent
molars are added as the jaw grows larger; the hindmost

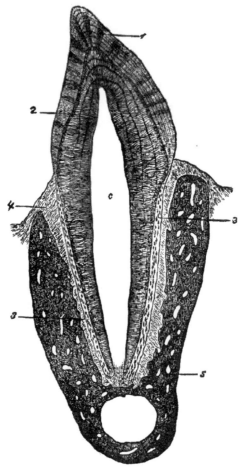

FIG. 29.—Section through a tooth still imbedded in its socket. 1, enamel; 2, dentine; 3, cement; 4, the gum; 5, the bone of the lower jaw; *c*, the pulp-cavity.

ones, often named the *wisdom-teeth*, do not appear until
about the twentieth year of life.

10. The Pulp of a Tooth.—If a tooth be broken open, a cavity (*c*, Fig. 29) will be found inside it. It is named the *pulp-cavity*, and during life is filled with a soft red very sensitive core, full of blood and nerves, named the *pulp*. At the tip of the fang, or of each fang, if the tooth has more than one, is a small aperture through which the nerves and blood enter. The pulp nourishes the tooth; on account of the nerves in it, it gives rise to great pain when exposed or inflamed. When a dentist speaks of destroying or removing the "nerve" of a tooth, he means the pulp.

11. The Hard Parts of a Tooth (Fig. 29) are made of three different materials. Surrounding the pulp-cavity is *dentine* or ivory: an elephant's tusk is made of dentine. Covering the ivory in the crown, is *enamel*, the tissue of the body which contains least animal matter. It is so hard that it will strike a spark with steel. Covering the dentine in the fang, is what has been named the *cement;* it is but a thin layer of bone under another name. The dentine is harder than bone, though not so hard as enamel.

12. Hygiene of the Teeth.—A great portion of the hard parts of a tooth consists of a very hard kind of chalk, and like chalk it is readily eaten away, or dissolved, by sour or *acid* liquids. The mouth should therefore be well washed after eating lemons or other sour things· and acid medicines should be sucked through a glass

10. What is the pulp of the tooth ? ' How do blood and nerves get into the pulp ? Use of the pulp. Why called the nerve ?
11. What is dentine? Enamel? Cement? How does dentine compare with enamel ?
12. What is the effect of acids on the teeth ? What precautions are therefore necessary for their preservation ? How may acids be made in the mouth ? What is said of decay of the teeth ?

tube, and swallowed after as little contact with the teeth as may be possible.

Many foods if kept in the warm moist mouth, decompose and give rise to acids: the teeth should therefore be thoroughly cleansed twice daily, with a soft toothbrush and tepid water. Finely powdered chalk or a little soap may be placed on the brush with advantage, as they counteract any acids which may be present. The enamel is not so easily attacked as the deeper parts of a tooth; but once the enamel is injured, the dentine is apt to decay rapidly. Small cavities in the enamel are not easily discovered unless they are on the outer side of the tooth. Remnants of food collect in them and, making acids, rapidly eat away the tooth. The teeth should therefore be thoroughly examined by a dentist two or three times a year, and all cavities filled.

13. The Tongue (Fig. 60) is endowed not only with a delicate sense of touch, but is the chief organ of the sense of taste. Being highly muscular and very movable, it also plays a great part in guiding food inside the mouth, so as to push it between the teeth until it is properly chewed, and then to drive it on into the pharynx to be swallowed. As an organ of taste, we shall study the tongue later (Chap. XXI.).

14. What a " Furred Tongue" Indicates.—In health the mucous membrane covering the tongue is moist, covered by little "fur" and, in childhood, of a bright red color. In adults, the natural color of the tongue is less red, except around the edges and at the tip. When any part of

13. Of what is the tongue the chief organ? What muscular work does it perform?
14. What is said of the covering of the tongue? Color? Indications of disordered digestion?

the alimentary canal farther on is out of order, the tongue is apt to be covered with a thick yellowish coating, and there is a "bad taste" in the mouth. This may in most cases be taken as a sign that there is something wrong with the stomach.

15. The Salivary Glands.—The liquid which moistens the mouth is named *saliva*. It consists of a slimy fluid, named *mucus*, made, or *secreted*, as we say in physiology, by the tiny glands of the mucous membrane, mixed with a more watery secretion made by three pairs of *salivary glands*. These glands lie outside the mouth, but pour their secretion into it through tubes or *ducts*. Two of the salivary glands are placed in front of the ears; their ducts open on the inside of the cheek opposite the second upper molar tooth. In the disease known as *mumps* they become greatly swollen. The other salivary glands lie between the halves of the lower jawbone. Their ducts open into the mouth beneath the tongue.

16. The Uses of Saliva are several. (1) It keeps the mouth moist and enables us to speak with comfort. This is well illustrated by the trouble from dryness of the mouth experienced by many young orators when they first try to speak in public. The dryness is due to the fact that nervous excitement for a time paralyzes the salivary glands and stops their secretion. (2) The saliva enables us to swallow dry food. A cracker when chewed would give rise merely to a heap of dust,

15 What is saliva? How made? Describe the position of the salivary glands. Where do the ducts of each pair open?
16. What is the first use of saliva? Illustrate. The second? Illustrate The third? Illustrate. The fourth?

impossible to swallow, if it were not moistened. This fact was made use of in the former East Indian rice-ordeal. All suspected persons were brought together and given parched rice to eat. The guilty individual, believing that his gods would bring his crime to light, usually had his salivary glands paralyzed by fear, and so could not secrete enough saliva to enable him to swallow the dry rice; while those with clear consciences had no difficulty. (3) Saliva, by dissolving many solid substances, enables us to taste them. Things in the solid state cannot be tasted, as you may easily discover by wiping your tongue dry and placing a piece of lump-sugar on it. Until a little moisture has come out and dissolved some of the sugar, no taste will be perceived. (4) Saliva turns starch, which is not itself nourishing, into sugar, which is.

17. Digestion in the Mouth.—By means of the teeth, the solid parts of our food are cut and crushed. At the same time, they are softened and made ready for swallowing by mixture with the saliva. Saliva also alters some nourishing substances in the food, and so changes them that instead of being insoluble they become readily soluble.

18. The Action of Saliva upon Starch.—Raw starch may be mixed with water, but will not dissolve in it. After a while, all the starch settles down from such a mixture. When starch is boiled in water, it swells up very much and mixes more thoroughly with the water than raw starch does, but still it does not dissolve. If

17. How is digestion carried on in the mouth? What are the uses of saliva?
18. What happens when starch is mixed with water? When boiled? What happens if you strain a solution of sugar and water?

you dissolve some salt or sugar in water, and pour the
solution into a bag made of three or four thicknesses
of very fine muslin, the salt or sugar will come through
just as freely as the water. But if you try the same
experiment with boiled starch, you will find that the
water comes through, but leaves most of the starch
behind it inside the bag. The tiny openings or pores
of the mucous membrane lining the alimentary canal,
through which the dissolved food has to pass when it is
absorbed into the blood, are far smaller than the holes in
the finest muslin; and starch, whether raw or boiled, could
not get through them. The saliva turns starch into sugar,
which dissolves rapidly and is very easily absorbed by
the mucous membrane. In this way bread and corn
and arrowroot and many other articles of diet which
contain much starch (p. 87) are enabled to nourish our
bodies.

19. Why Food should be well Masticated.—Some persons eat as if all that their teeth and mouth had to do
was to bite and swallow : they seem to believe that their
stomachs are like the gizzard of a bird, constructed to
crush and grind. Nature having provided man with
teeth, has given him no gizzard : the human stomach
will certainly get out of order if it is frequently called
upon to do the work of one. Our molar teeth are so

Of starch? How may the pores of the mucous membrane of the
alimentary canal be compared to muslin? How does the action of
saliva enable starch to get through these pores? Why could we not
digest bread, corn, arrowroot, and like food without saliva?

19. What duty besides biting and swallowing have the teeth in
connection with digestion? Where do fowls crush hard food? What
is the consequence if we eat as if we had gizzards? What is the evi-
dent duty of our molars? How does chewing affect the salivary
glands?

clearly fitted to break up our food into small pieces that there can be no doubt as to what their use is.

The *chewing* or *mastication* of food also causes a greater flow of saliva. When we are not eating, the salivary glands secrete little; but as soon as we commence to chew, they begin to be more active. If food be swallowed hastily, it is not mixed with sufficient saliva, and in consequence, its starchy parts are imperfectly digested.

CHAPTER XI.

DIGESTION, CONCLUDED.

1. The Pharynx (Fig. 30) is a muscular bag lined by mucous membrane; it opens at its lower end into the gullet, *b*. Not only our food, but also the air which we breathe, has to pass through the pharynx, for into its upper portion, above the level of the palate, *l, f*, the inner ends of the nostril-chambers open. Under the soft palate, *f*, is the aperture through which food is sent from the mouth; and, lower still, another opening, behind the root of the tongue, through which air enters the passage, *c*, which transmits it to the lungs.

2. Swallowing or Deglutition is the process of sending food or drink from the mouth to the stomach. The liquid, or the mass of chewed solid food, is collected on the upper surface of the tongue, and then pushed into the pharynx. As soon as it has left the mouth, the opening between mouth and pharynx is closed, to prevent its return. At the same instant the soft palate is raised, so as to separate the upper from the lower portion of the pharynx: in this way the food is prevented from getting into the nose. The lid, *e*, named

1. What is the pharynx? What besides food passes through it? What opens into it above the palate? Below? Behind the root of the tongue?
2. What is deglutition? How is food sent from the mouth to the pharynx? How is its return prevented? How is it kept from getting

the *epiglottis*, which overhangs the aperture leading to the windpipe, *c*, is also shut down. Therefore, when the muscles of the pharynx contract and press on the food, the only way it can go is into the gullet, *b*. Occasionally a morsel "goes the wrong way," and gets into the air-passage, causing a fit of coughing which drives it back into the pharynx. The things which we swallow are hurried through the pharynx very fast, so as to get it clear, and enable us to breathe again.

3. The Passage of Food and Drink along the Gullet or œsophagus, is slow. A mouthful of food or drink when it has entered the œsophagus does not drop down that tube into the stomach, like a brick falling down a chimney, but is seized by the muscular rings in the coat of the gullet, which contract one after another and push it along.

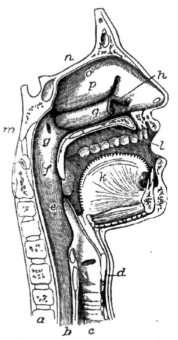

Fig. 30.—The mouth, nose, and pharynx, with the commencement of the gullet and larynx, as exposed by a section, a little to the left of the middle of the head. *a*, vertebral column; *b*, gullet; *c*, windpipe; *d*, larynx; *e*, epiglottis; *f*, soft palate; *g*, opening of Eustachian tube; the letters *e, f, g* are placed in the pharynx; *k*, tongue; *l*, hard palate; *m*, the sphenoid bone on the base of the skull; *n*, the fore part of the cranial cavity; *o, p, q*, the turbinate bones of the outer side of the left nostril-chamber.

into the nose? Into the windpipe? What does it enter when forced out of the pharynx? What is meant when a morsel of food is said to have gone the wrong way? Why is food sent quickly through the pharynx?

3. How do food and drink pass along the gullet? Illustrate.

For this reason, horses and many other animals are able to swallow, although they usually eat with their mouths much lower than their stomachs; and jugglers are able to drink a glass of water while standing on the head.

4. The Stomach (Fig. 31) is a dilated portion of the alimentary canal, which lies at the lower end of the œsophagus, in the upper part of the abdomen, rather more on the left than the right side of the body (see Fig. 2). Outside its lining mucous membrane, is a thick muscular coat.

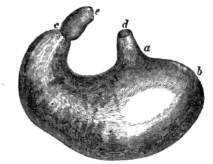

FIG. 31.—The stomach viewed from the front. *d*, lower end of the gullet; *a*, position of the cardiac aperture; *b*, the fundus; *c*, the pylorus; *e*, the first part of the small intestine.

The gullet, *d*, opens into the upper side of the stomach by an aperture named the *cardiac orifice.* The right end of the stomach gradually narrows to the commencement of the small intestine, *e.* The place, *c*, where stomach and intestine meet is named the *pylorus,* and the opening which places their cavities in communication is the *pyloric orifice.* When moderately distended, the stomach contains about three pints.

4. Position of the stomach? What is outside its mucous membrane? How and where does the gullet enter it? How does the stomach join the small intestines? What is the pylorus? The pyloric orifice? Capacity of the stomach?

5. The Gastric Juice.—The mucous membrane of the stomach is almost entirely made up of thousands of tiny glands, placed side by side nearly as close as they can be packed. The liquid which these glands make is poured into the stomach, and is known as the *gastric juice*. If you imagine a piece of honeycomb reduced very much in size, and that its cells answer to the glands, you will have a fair idea of how the glands lie in the mucous membrane of the stomach. To complete the resemblance, each cell would have to be open at one end, and through this opening to pour its honey on the surface of the comb, and to keep on making honey to take the place of that it had emptied out. The liquid, too, would have to be much thinner than honey, and sour or *acid*, instead of sweet.

6. Digestion in the Stomach. — When the healthy stomach is empty, its mucous membrane is something like grayish-pink velvet and its glands make hardly any gastric juice. As soon as food is swallowed, a great deal of blood flows to the mucous membrane, and it becomes red. At the same time, its glands secrete abundantly, and, all over the surface, gastric juice trickles out, like sweat on the skin of a person perspiring profusely. These facts were first observed many years ago, on a Canadian hunter, named Alexis St. Martin, who, as a result of a gunshot wound, had a small opening from the surface of his abdomen into his stomach. Through this opening, what was going on inside his stomach could

5. Of what is the mucous membrane of the stomach chiefly made up? What is the gastric juice? Illustrate.
6 What is the state of the stomach when empty? After swallowing food? How were these facts first observed? What are the chief foods acted on in the stomach? How are they changed?

be watched. Since then the careful observations made by his physician have been confirmed by the study of several similar cases.

The chief kinds of foods acted on in the stomach, are of albuminous nature (p. 85), lean meat, white of egg, cheese, the gluten of bread, and so forth. They are turned into a condition in which they can be dissolved and absorbed.

7. **The Muscular Coat of the Stomach** (Fig. 19) performs two duties: first, it thoroughly mixes our food with the gastric juice; and next, it drives it on into the intestine. The pyloric orifice (*c*, Fig. 31) is narrow, and surrounding it is a thick ring of muscle which keeps the passage closed, for an hour or more after eating. During this time, the muscles of the stomach contracting, now in one direction and now in another, keep its contents in constant motion and bring every part of the food into contact with the gastric juice.

When the digestive process has gone on until some food is ready to enter the intestine, the muscle around the pylorus relaxes a little from time to time; thus some liquified food is passed through the opening. When all the things which can be dissolved in the stomach have been passed on, the pyloric orifice opens wider and lets solid indigestible things get through. In this way buttons, coins, cherry-stones, and other such things which may have been swallowed reach the

---

7. What are the duties of the muscular coat of the stomach? What surrounds the pyloric orifice? Its use? What happens while the passage is closed? What occurs when some food has been prepared to enter the intestine? What occurs when the stomach has done all it can towards digesting its contents? How soon after an ordinary meal is the stomach empty?

bowels, to travel through them and, in fortunate cases, be sent out of the body, along with indigestible portions of the food. In health the stomach is completely emp-

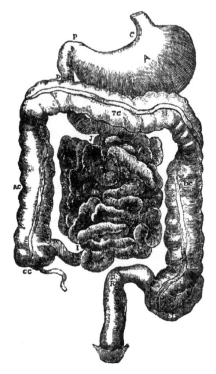

FIG. 32.—The abdominal portions of the alimentary canal. *A*, stomach; *C*, lower end of gullet ; *P*, pylorus ; *D*, *J*, *I*, portions of the small intestine, named respectively duodenum, jejunum, and ileum ; *CC*, *AC*, *TC*, *DC*, *SF*, *R*, portions of the large intestine, named respectively the cæcum, ascending colon, transverse colon, descending colon, sigmoid flexure, and rectum.

tied in from two and a half to three and a half hours after an ordinary meal.

8. **The Small Intestine** (*D*, *J*, *I*, Fig. 32), commenc-

8. Commencement, course, and ending of small intestine ? Length

ing at the pylorus, ends after many windings, by joining the large. In an adult it is about twenty feet long
and an inch and a half wide. Imbedded in its mucous
membrane, are myriads of tiny glands, much like those of
the stomach in shape and arrangement, but preparing
a digestive liquid very different from the.gastric juice.
This liquid is mixed with the food, as it is slowly
driven along by the muscles in the coat of the intestine.
In addition, two large glands, the *liver* and the *pancreas*,
pour their secretion into the small intestine near its
upper end.

9. **The Liver** is by far the largest gland in the body.
It is placed in the upper part of the abdomen on the
right side (Fig. 2, *le, le'*), close under the diaphragm.
The secretion of the liver is named *bile* or *gall*. When
no food is being digested in the intestine, the bile collects in a pear-shaped bag, the *gall-bladder*, which lies
under the liver. As soon as food is sent on from the
stomach, the gall-bladder empties bile upon it through a
tube or duct which opens into the intestine about opposite *D*, Fig. 32.

Fresh human bile is a yellow-brown liquid. It is much
like weak lye in some of its properties; and ox-bile or
ox-gall is occasionally used by housekeepers instead of
lye, for cleansing purposes; to dissolve and remove
grease-spots. One chief use of bile is to aid in digesting the oily and fatty parts of our food.

and width? The glands of its mucous membrane? Their secretion? What other glands pour secretion into the small intestine?
Where?
9. What is said of the liver? Its position? Name of its secretion? Where stored when not needed? How disposed of when
food enters the intestine? Color of bile? Why sometimes used in
housekeeping? Use of bile in digestion?

10. The Pancreas lies along the lower side of the stomach. Its duct opens into the small intestine at the same place as the bile-duct. The secretion of the pancreas is named *pancreatic juice*. It is much like saliva in appearance, being transparent and colorless. The pancreatic juice is one of the most important digestive liquids. It acts upon starch as saliva does, turning it into sugar; it dissolves albuminous matters, thus completing the action of the gastric juice; and, more powerfully than bile, it promotes the absorption of fatty food.

11. Digestion in the Small Intestine.—The soft half-digested food-mass which enters the intestine from the stomach, is named *chyme*. It is at once mixed with bile and pancreatic juice, and then, as it slowly passes along, has the secretion of the innumerable little glands of the mucous membrane of the small intestine added to it.

The result of the combined action of these liquids is that any starch which escaped digestion in the mouth, is turned into sugar; any albuminous substances which had not been fully dissolved in the stomach are finally digested; fats, which are not acted upon at all by either saliva or gastric juice, are prepared for absorption. Digestion in the small intestine is on the whole more important than that which takes place in the mouth or stomach. The product of intestinal digestion is a highly nutritious creamy liquid containing all the nourishing matters of the food. This liquid is named *chyle*. It is

10 Position of the pancreas? Opening of its duct? Name of its secretion? Appearance? Value? Action on various foodstuffs?

11 What is chyme? What is at once mixed with it in the intestine? What afterwards? How do these liquids act on starch? On albumens? On fats? Where does the most important part of digestion occur? What is said of the liquid produced by digestion in the small intestine? Its name? For what is it ready?

ready to be taken up into the blood and carried to every organ.

12. Absorption from the Small Intestine.—As the chyle passes along, it is gradually absorbed by the mucous membrane, which is specially adapted to fulfil this duty. Instead of being nearly smooth like the mucous membrane lining the mouth, it is raised up into numerous folds (Fig. 33) which greatly increase the extent of its surface; and thus it is enabled to absorb more and quicker than if it was stretched smooth and flat. The

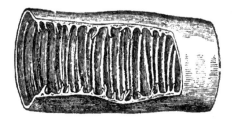

Fig. 33.—A portion of the small intestine opened to show the folds of its mucous membrane.

pockets or hollows between the folds also have their use. The chyle collects in them and is thus prevented from passing along faster than it can be absorbed.

13. The Villi of the Small Intestine.—All over the mucous membrane of the small intestine, both on its folds and between them, are tiny elevations, which stand up like the pile on velvet. Each elevation is a *villus*, and, small though it is, contains two sets of tubes or *vessels*. One

12. For what is the mucous membrane of the small intestine specially adapted? How is the extent of the surface of its mucous membrane increased? Use of these folds? Of the hollows between them?

13. Describe the surface of the mucous membrane of the small intestine. What is a villus? What does each villus contain? Use of the villi? Where does the chyle absorbed by them go?

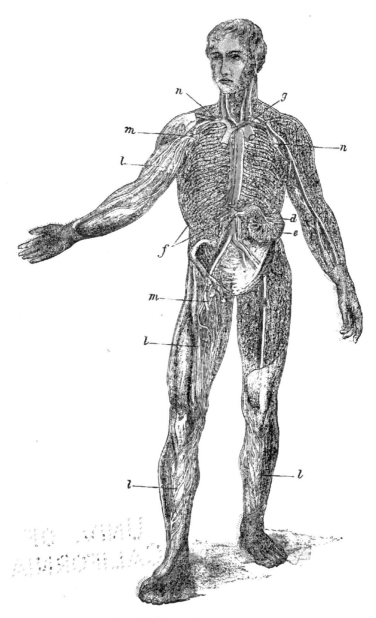

PLATE III.—A GENERAL VIEW OF THE LYMPHATICS OR ABSORBENTS.

A General View of the Lymphatic or Absorbent System
of Vessels.

e, A portion of the small intestine from which lacteals or chyle-conveying vessels, *d*, proceed their origin within the villi may be seen magnified in fig. 34; *f*, the duct called thoracic, into which the lacteals open. This duct passes up the back of the chest, and opens into the great veins at *g*, on the left side of the neck: here the chyle mingles with the venous blood In the right upper and lower limbs the superficial lymphatic vessels, *l l l l*, which lie beneath the skin, are represented. In the left upper and lower limbs the deep lymphatic vessels which accompany the deep blood-vessels are shown. The lymphatic vessels of the lower limbs join the thoracic duct at the spot where the lacteals open into it: those from the left upper limb and from the left side of the head and neck open into that duct at the root of the neck. The lymphatics from the right upper limb and from the right side of the head and neck join the great veins at *n*. *m m*, enlargements called lymphatic glands, situated in the course of the lymphatic vessels. These vessels convey a fluid called lymph, which mingles with the blood in the great veins.

set (*d*, Fig. 34) carry blood; the other, *b*, a watery liquid named *lymph*. The villi act like little roots or suckers, and the chyle which they absorb, goes, some of it into the blood at once, and some into the lymph-vessels.

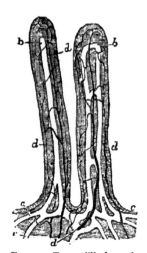

14. The Lacteals.—Lymph-vessels, like blood-vessels, are found in nearly every part of the body. Another name by which they are known is the *absorbents*. Most lymph-vessels contain only a thin colorless liquid, the *lymph*. But when chyle is being absorbed, the lymph-vessels of the small intestine take up so much of it that their contents become white and milky-looking. Hence they have been named the *lacteals*,

FIG. 34.—Two villi from the inside of the small intestine, magnified eighty times; *d*, *d*, blood-vessels; *b* and *c*, lymphatic vessels or lacteals.

from a Latin word (*lac*) meaning milk. The lacteals finally pour the chyle which they have taken up, into a tube, named the *thoracic duct*, which runs up to the bottom of the neck and there opens into a large blood-vessel.

15. The Large Intestine (*CC, AC, TC, DC, SF, R*, Fig. 32) is two or three times as wide as the small, but only about five feet long. In it the absorption of the nourishing part of the food is completed, and towards its lower

14. Where are lymph-vessels found? Another name for them? What do most contain? What do those of the intestine become filled with during digestion? What name has been given them? What do they do with the chyle? Where does the thoracic duct pour its contents into the blood?
15. Size of the large intestine? Uses?

end the indigestible residue collects, ready to be expelled from the body.

16. Summary.—When digestion and absorption are completed, all the useful portions of a meal have at last been mixed with the blood. Some of them, as water, were ready for absorption without undergoing any change; all we had to do was to swallow them, and the coats of the stomach took them up at once, if there was not too much of them. Others, as a pinch of salt or a lump of sugar, were ready to dissolve at once. Still others, like the lean of meat, and starchy foods, had to be changed by the digestive liquids before they could be dissolved.

Some were changed by saliva, some by the gastric juice, others by the liquids of the intestines; but sooner or later, in mouth or stomach or bowels, they were made ready for absorption.

Some of the nutritive liquid was absorbed by the blood-vessels of the stomach; more by the blood-vessels of the intestinal villi; still more by the lacteals. What little may still have been left, was sucked up into the blood- and lymph-vessels of the large intestine. But no matter where it was absorbed, or by what vessels, it finally reaches the blood, and supplies it with water and minerals and albumens and fats and sugar, to be carried to every organ.

16. What has happened when the digestion and absorption of a meal are completed? Name a food-stuff absorbed without change. One which has to be simply dissolved. Some which had to be changed by the digestive juices before they could be absorbed. Name the liquids used in changing them. Name the vessels concerned in the absorption With what does the absorbed liquid supply the blood?

APPENDIX TO CHAPTER XI.

The main points in the anatomy of the alimentary canal may be easily studied on a kitten, puppy, or rat   Superfluous kittens and puppies have so often to be drowned, that no unnecessary taking of life is called for.   The animal may be more mercifully killed by shutting it up in a small tight box, for ten minutes, along with a small sponge soaked in chloroform   A tin cracker-box does very well.

Cut away, with strong scissors, the front of the chest and abdomen of the dead animal, taking care not to injure the contents of those cavities.   Dissect off the skin on the front and sides of the neck.   Remove the larynx, trachea, lungs, and heart.

The *gullet*, a slender muscular tube, will now be exposed in the neck; trace it through the chest; note the relative positions of the abdominal viscera as now exposed, before displacing any of them; then turning the liver up out of the way, follow the gullet in the abdomen until it ends in the stomach.

Note the form of the latter organ; its projection (*fundus*) to the left of the entry of the gullet, its *great* and *small curvatures;* its narrower *pyloric portion* on the right, from which the small intestine proceeds.   Attached to the stomach, and hanging down over the other abdominal viscera, notice a thin membrane, the *omentum.*

Follow and unravel the coils of the small intestine, spreading out as far as possible the delicate membrane (*mesentery*) which slings it.   In the mesentery are numerous bands of fat, running in which will be seen blood-vessels and lacteals.

The termination of the small intestine by opening into the large. Observe the *cæcum* or blind end of the latter, projecting on one side of the point of entry of the small intestine; on the other side follow the large intestine until it ends at the anal aperture, cutting away the front of the pelvis to follow its terminal portion (*rectum*).   The portion between the cæcum and the rectum is the *colon.*

Spread out the portion of the mesentery lying in the concavity of the first coil (*duodenum*) of the small intestine; in it will be seen a glandular mass   the *pancreas.*

Observe the *portal vein* entering the under side of the liver by several branches   Alongside it will be seen the *gall-duct*, formed by the union of two main branches, and proceeding, as a slender tube, to open into the duodenum a short way from the pyloric orifice of the stomach.   In a kitten or puppy the *gall-bladder* will be seen on the under side of the right half of the liver.

Note the *spleen·* an elongated red body lying in the mesentery, behind and to the left of the stomach.

Divide the gullet at the top of the neck, and the rectum close to the anus. and, severing mesenteric bands, etc., by which intermediate portions of the alimentary canal are fixed, remove the whole tube; then cutting away the mesentery, spread it out at full length, and note the relative length and diameter of its various parts; and that the small intestine forms by far its longest portion.

Open the stomach, note, in the rat, that the *mucous membrane* lining the fundus is thin and smooth, and is sharply marked off from the thick corrugated mucous membrane lining the rest of the organ. This is not the case in the stomach of dog, cat, or man. Pass probes through the *cardiac orifice* into the gullet and through the *pyloric orifice* into the duodenum.

CHAPTER XII.

HYGIENE OF THE DIGESTIVE ORGANS.

1. Why Care of our Digestion is an Important Duty.—
When the digestive organs are all in good working
condition, appetite is healthy, our meals are enjoyed, the
temper is cheerful, the body vigorous and well nourished, and life pleasant. If the stomach, liver, or intestines fail in their duty, the picture is reversed. Appetite
is wanting, the brain and muscles are ill nourished, everything is regarded from a gloomy point of view, work of
any kind is a burden, and life a weariness. It is plain
from this, that in most cases, the man with a good digestion and, in consequence, a well-nourished body and
cheerful active mind, has much better chances for success in life than one whose energies are weighed down
by bad digestion, and consequent ill heath.

Not merely from its bearing on our own personal
happiness and welfare, but as a matter of duty towards
others, it is of primary importance to preserve the
digestive organs in a healthy natural state. A man
with good digestion is pretty sure to be an agreeable
and encouraging friend or companion; while the chances
are that one who digests badly, will be irritable, depressed, and disagreeable.

1. How does a good digestion influence our health and happiness?
A bad? Our chances for success in life? Why is it a duty towards
others to try and maintain our digestive organs in good condition?

2. Dyspepsia is the name commonly used to indicate difficult or painful digestion. It may take many forms and be due to imperfect action of different organs. In certain cases, it is, no doubt, unavoidable; some unfortunate people have weak stomachs, or sluggish livers, as others have feeble muscles or poor· eyes, from causes beyond their control. But in the great majority of cases, dyspepsia is due to some imprudence in conduct. Its most frequent cause is unwise eating and drinking; but mental overwork, neglect of muscular exercise, lack of fresh air, late hours, and improper clothing, all play their part in various cases. Probably not more than one person out of five of those who live in towns or cities, reaches the age of forty, without suffering from some form of dyspepsia, which might have been avoided by wiser habits during early life. Once it has made its appearance, dyspepsia is extremely difficult to get rid of. How best to avoid it, is therefore a very important branch of hygiene.

3. The Intervals between beginning Meals should be not less than four hours, five is better; except in the case of young children and invalids, who require food more often, and in small quantities at a time. As we have learned (p. 115), the stomach is only emptied about two and a half or three hours after an ordinary meal. It should have some rest before being set again to work. During this rest, it collects in its glands material for making a fresh supply of gastric juice. Eating between

2. What is dyspepsia? Why not always avoidable? What is said concerning it in the majority of cases? Most frequent cause? Other causes? What is said concerning dyspepsia in those who live in cities? Concerning ease of cure?

3. Proper time between meals? Exceptions? What does the stomach collect during rest? Why is eating between meals injurious?

meals keeps the stomach at work all the time, and it is not ready to do its duty properly when the meal-time comes.

4. Meals should be taken at Regular Hours.—Three meals a day are sufficient, and many persons do better with two. Their regularity is of more importance than their number. The stomach, like the rest of our organs, soon forms habits, and only works with comfort when they are not interfered with. A little before the usual time of eating, we begin to feel an appetite, which gradually increases to a pleasant degree of hunger. If a man keeps at his work instead of heeding this hint, his whole system becomes run down from want of nourishment. His stomach itself is unable to secrete properly, and when at last he sits down to eat, utterly tired out, he has no appetite, and his meal is probably followed by a fit of indigestion.

A heavy meal should not be eaten within two hours of going to bed. The presence of much food in the stomach, is very apt to cause troubled sleep.

5. Meals should be Eaten Slowly and with Pleasant Surroundings.—The dinner-table should be the scene of a cheerful gathering, and merry talk. To bolt a meal in gloomy silence, thinking of one's work or worries, is not only bad manners, but bad hygiene. Talking of unpleasant things is often necessary, but should always be put off until an hour or two after dinner, when one is in much better condition to meet annoyances.

4. Proper number of meals ? Why should they be taken at regular times ? What happens if a man allows his business to postpone a meal much beyond the proper time ? What is said concerning eating before bedtime ?

5. What is said concerning the dinner-table ? About eating while thinking of work or worry ? About talking of unpleasant things ?

6. Some Rest should be taken before Eating.—When the mind or body is greatly tired, the digestive organs will not act properly. A man engaged in any laborious business should therefore take his dinner after he has finished his day's work, and had at least half an hour's rest. If he come straight from his office to the table, and return there directly after eating, he will, in the long-run, injure his health.

The day's work of a child should finish early, and be followed, after an hour's recreation, by dinner. If there is an afternoon session at school, the work should be of a kind calling for little mental effort; for example, writing or drawing.

7. Food should be Eaten Slowly.—This ensures its proper mastication and thorough mixture with the saliva (p. 106). Thus the work of the stomach and other digestive organs is lightened.

Moreover, rapid eating is very apt to lead to over-eating. Too much food is swallowed before enough has been absorbed to lead to diminution of the sensation of hunger.

8. The Proper Amount of Food varies with age, work, and climate. A person who has done growing, needs only enough to make good his daily waste, while a child should have something over, to supply materials for growth. In warm weather, less food is required than in cold, because less material has to be oxidized in the

6. Why is it unwise to eat when very tired? When should a man whose business is fatiguing dine? Why? What is said concerning the day's work of a child? Of afternoon session at school?

7. Why should we eat slowly? How is it that rapid eating may lead to over-eating?

8. Conditions affecting proper amount of food? Need of an adult? Of a child? Influence of weather? What is a safe guide? What is

body (p. 82) to keep up the animal heat. If food be eaten slowly, the natural appetite is, in health, a safe guide. Those who injure themselves by over-eating, are not the workers who come to their meals hungry, but the indolent who, having little appetite, stimulate their palates by highly flavored food, to enable them to eat what they have not earned, and their bodies do not want.

An over-distended stomach is not merely injured itself, but interferes with the heart and lungs. It pushes the diaphragm up against them and impedes their movements. Hence result feelings of oppression in the chest, shortness of breath, and faintness. *Palpitation of the heart* may also be produced : it is frequent in that kind of dyspepsia which is accompanied by accumulation of gas in the stomach.

9. A Proper Diet contains both Animal and Vegetable Foods.—The teeth of a purely flesh-eating animal, as a tiger, are constructed only for tearing and cutting. Those of a vegetable-eater, as a cow, are very broad and constructed for grinding, except a few in front for cropping grass. Man's teeth are half-way between the purely flesh-eating and purely vegetable-eating kind. Their structure shows that our proper food is both animal and vegetable.

This is also proved by the fact that some of the digestive liquids found in the human alimentary canal are, like the saliva (p. 106), especially fitted to digest

said concerning those who injure themselves by over-eating? Consequences of an over-distended stomach? What is said concerning palpitation of the heart? `

9 Teeth of a purely flesh-eating animal? Of a vegetable-eater? Of man? What does their structure show? How is this also proved? When should less animal food be taken?

starch, which is the chief food-stuff in vegetables; while others, as gastric juice (p. 113), are adapted to digest albumen, which is scarce in vegetables but abundant in most animal foods.

In warm weather, when the body easily keeps up its animal heat, it is not well to eat much animal food.

10. Drinking much Water during a Meal is Injurious.—The gastric juice acts well only in warmth, and a glass of cold water cools the stomach very much, and for a considerable time. Cold, also, drives the blood out of the mucous membrane, just as it would out of the skin. This stops or diminishes the action of the glands, which can only pour out abundant gastric juice when they are richly supplied with blood. If water be taken slowly, and only a few mouthfuls at a time, a much smaller quantity will satisfy the thirst, than if a glassful be taken at a draught. Also, the stomach is not enough cooled, at any moment, to interfere with digestion. Moreover there is a limit to the amount of water which the coats of the stomach will quickly absorb. Any more than that will be left over and make the gastric juice too weak to work well.

About a single glass with a meal in cool weather, and two glasses in warm, is a proper quantity.

In warm weather or after heating exercise it is well to assuage thirst at least half an hour before going to table, so that the water may be absorbed before the stomach is called upon to digest.

10 What is necessary in order that the gastric juice may act well? What is the effect of drinking a glass of ice-water? How does cold act on the blood in the mucous membrane? What is the consequence as regards the glands of the stomach? Why should water be drunk slowly? The proper quantity with a meal? Why assuage thirst some time before eating?

11 Exposure of the Skin to Cold often causes Disease of the Digestive Organs.—Every one knows that eating certain things, as unripe fruit, is apt to cause colic and diarrhœa. But a more frequent cause of these complaints is insufficient clothing A man goes out on a summer morning with no cotton or woollen under-garments, gets very hot at his day's work, comes home tired, and not able to withstand any extra strain on his organs. By way of becoming cooled and refreshed, he sits, with no extra clothing, in a draught. This chills the skin unduly, and the blood driven from it (p. 78) collects in internal organs in excessive quantity. A common result is that the person feels chilly and uncomfortable before going to bed, and is awakened in the night suffering from colic and diarrhœa. Diarrhœa is nearly always due to excessive secretion by the mucous membrane lining the bowels. This being inflamed secretes excess of waterly liquid, like the membrane lining the nose in a "cold in the head." In both cases watery matter that ought to have been carried off by the skin is driven back to the interior. Draughts should always be avoided, but especially if the underclothing be damp with perspiration Its rapid evaporation, by cooling the skin (p. 67) very fast, much increases the danger. If, in such circumstances, you have to sit in a current of air, throw an extra wrap around you, at least until the clothes next the skin have become dry. You will thus gain your end of getting to feel cool, nearly as soon as

11. What is apt to result from eating unripe fruit? What is a more common cause of colic? Example? To what is diarrhœa directly due? When should draughts be especially avoided? Why? What should be done if you cannot avoid sitting in a draught when perspiring?

if you neglect this precaution, and with much less danger.

12. The Large Intestine should be trained to empty itself once a day, at a regular hour. Neglect of this leads to the retention of injurious substances in the body.

13. The Action of Alcoholic Drinks on the Digestive Organs is such as to frequently cause disease. In some cases this is due to a general slow poisoning of all parts of the body, enfeebling it and rendering impossible the healthy activity of any organ. Two digestive organs are, however, especially apt to be attacked by alcoholic disease. They are the *stomach* and the *liver*.

14. The Action of Alcohol on the Stomach is first to cause its mucous membrane to become overgorged with blood, or, in medical language, *congested.* If the dose be not very large or soon repeated, this passes off, as the alcohol is absorbed and carried off by the blood to work mischief elsewhere.

But repeated tippling keeps the stomach in this congested state. Instead of being allowed a period of rest before every meal, it is kept excited all the time and very soon becomes inflamed. Appetite is lost or replaced by nausea; the stomach, accustomed to the powerful alcoholic stimulant, does not pour out gastric juice when less stimulating food enters it, and dyspepsia is the consequence.

12. What is said concerning the large intestine?
13. Action of alcoholic drinks on the digestive organs? To what sometimes due? What digestive organs are especially apt to be injured by alcohol?
14 First action of alcohol on the stomach? What follows if a fresh dose be not soon taken? Effect of repeated tippling?

For a time, a person in this condition finds that another glass of spirits or wine creates appetite and, by exciting the stomach to secrete, promotes digestion. So he falls daily more and more into the habit of drinking. The consequence is that the stomach at last ceases to be able to make gastric juice at all. The usual glass now fails to produce appetite, and food if swallowed is not digested. Unless a very strong effort be made to break the habit, and skilful treatment be long employed to get the stomach back into a healthy state, a man in this condition is sure to die a drunkard.

15. A Single Large Dose of Alcohol or of a drink containing it frequently irritates the stomach so much as to cause vomiting. This has saved the lives of many foolish people. Occasionally a very large dose paralyzes the stomach for a while, so that it does not absorb; this is sometimes seen when a man, for a bet, undertakes to drink a bottle of whiskey in a few minutes. If his stomach does not reject it, he often appears unaffected for half an hour or so: then suddenly falls down drunk, and often dies in a short time. This occurs when the stomach, having begun to recover from the first shock, suddenly commences to absorb the alcohol.

16. The Action of Alcohol on the Liver.—All the blood which flows through the mucous membrane of the stomach goes straight to the liver, before it is carried to any other organ of the body. This blood of course takes

Explain. Consequences? Why is a tippler apt to fall more into the habit? Consequences?

15. Usual action of a large dose of alcohol on the stomach? Occasional result? Example?

16 Where does blood leaving the stomach go next? What might we expect as regards the result on the liver of alcohol-drinking?

with it whatever it has absorbed from the stomach. It
is, therefore, not strange that the liver often becomes
diseased from a man's taking alcoholic drinks. They
cause a great overgrowth of the connective tissue (p. 13)
of the liver, giving rise to what is known as *fibrous degeneration.* The true liver-substance is crushed and killed,
and what remains is a shrunken, hard rough mass, well
known to physicians as "hob-nailed," or "gin-drinker's
liver."

A liver in this condition cannot, of course, secrete bile
properly, and thus digestion is imperfect.

In still another way, the nourishment of the body is
very seriously affected. Besides making bile, the liver
has another duty. This is to take up from the blood
the sugar (much of it made from vegetable starch, p. 108)
which the blood has absorbed from the stomach or intestines, and turn this sugar back into a kind of *animal
starch,* which is more fitted to nourish the muscles and
several other organs. This animal starch is then returned
to the blood to be carried over the body. A diseased
liver cannot perform this duty, and many organs are in
consequence ill nourished.

What is fibrous degeneration? Gin-drinker's liver? Result as regards digestion? Another use of the liver besides making bile?
What becomes of the animal starch made in the liver? Result when
liver is diseased?

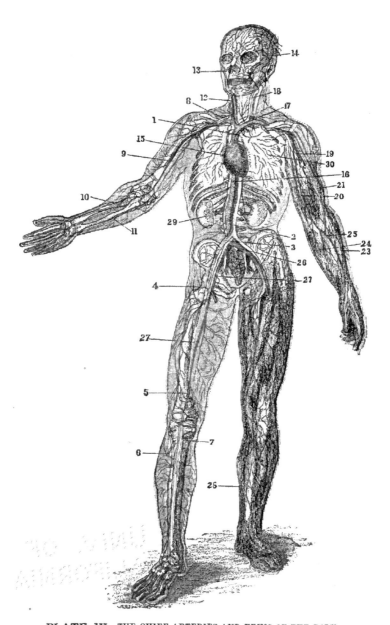

PLATE IV.—THE CHIEF ARTERIES AND VEINS OF THE BODY.

EXPLANATION OF PLATE IV.

The Circulatory Organs.

The arteries (except the pulmonary) and the left side of the heart are colored red; the veins, (except the pulmonary) and the right half of the heart blue: on the limbs of the left side the arteries are omitted and only the superficial veins are shown.

1. Aorta, near its origin from the left ventricle of the heart.
2. Lower end of aorta.
3 Iliac artery
4. Femoral artery.
5. Popliteal artery; the continuation of the femoral which passes behind the knee-joint.
6, 7 The main trunks (anterior and posterior tibial arteries) into which the popliteal divides).
8. Subclavian artery.
9. Brachial artery.
10. Radial artery.
11. Ulnar artery.
12. Common carotid artery.
13. Facial artery.
14. Temporal artery.
15. Right side of Heart, with superior vena cava joining it above, and inferior vena cava (16) passing up to it from below.
17. Innominate vein, formed by the union of subclavian and jugular veins. The right and left innominate veins unite to form the superior cava.
18. Left internal jugular vein.
19. Axillary vein.
20. Basilic vein.
23. Radial vein.
24. Ulnar vein.
25. Median vein.
26. Iliac vein.
27. Femoral vein
28. Long saphenous vein
29. The kidney; attached to it are seen the renal artery and vein.
80. Branches of the pulmonary arteries and veins in the lung.

CHAPTER XIII.

THE CIRCULATION OF THE BLOOD.

1. The Circulation.—Blood is not allowed to lie at rest in any part of the body. It is kept all the time moving round and round, from organ to organ, through a set of tubes, *the blood-vessels*. This regular flow of the blood is named the *circulation*.

2. The Use of the Circulation.—If blood which had been enriched by the absorption of nourishment from the alimentary canal should remain stationary, the muscles and brain would be starved. If blood in the skin and that in deeper parts did not change places, the skin would become too cold, and the inside of the body too hot (p. 78). If blood in the muscles were not kept moving on, and fresh blood taking its place, it would soon become so loaded with waste matters from the working muscles that it would poison them. It has to be carried off to organs (the lungs, Chap. XV., and kidneys, Chap. XVII.) in which these injurious matters are separated from it, and it is thus made again ready for use. By means of the circulation, then, the blood flows through every organ in turn; here becoming rich

1. What is the circulation of the blood?
2. How might blood have nourishment in it and yet brain and muscles be starved? What would happen, as regards the heat of the body, if the blood were not kept flowing through the skin? How might the blood in a muscle poison it? What organs separate waste substances from the blood? What is accomplished by the circulation?

in foods, there feeding the organs; here warmed, and there cooled; here loaded with wastes, and there purified. Thus by the flowing blood, every part is cared for.

3. The Organs of Circulation are the *heart*, the *arteries*, the *capillaries*, and the *veins*. The heart is a hollow muscle which squeezes the blood on, and keeps it moving. The arteries carry blood from the heart and distribute it over the body. The capillaries are very fine tubes with very thin walls, into which the arteries of every organ pour their blood. The veins take up blood from the capillaries and carry it back to the heart.

4. The Blood, as every one knows, is a red liquid which is very widely distributed over the body, since it flows from any part of the surface when the skin is cut through. There are very few portions of the body into which blood is not carried. The outer layer of the skin (Chap. VI.), the hairs and nails, the hard parts of the teeth, and most cartilages contain no blood; these *non-vascular* tissues are nourished by liquid which soaks through the walls of blood-vessels in neighboring parts.

5. Arterial and Venous Blood.—Although all blood is red, it is not all the same tint of red. In nearly all arteries, the blood, just sent out of the heart, is bright scarlet; such blood is named *arterial blood.* In nearly all veins, the blood, which has just flowed through the capillaries of some organ, is of a dark purple-red color; such blood is named *venous blood.*

3. Name the organs of the circulation. What is the heart? What are the arteries? The capillaries? The veins?

4. Distribution of blood in the body? Portions which get no blood? How are the non-vascular tissues nourished?

5. What is arterial blood? Venous?

6. The Corpuscles of the Blood.—Fresh-drawn blood is, to the unaided eye, a uniform red liquid. But a microscope shows it to consist of a colorless liquid, the *blood-plasma*, in which float vast numbers of tiny solid particles, named the *blood-corpuscles* (Fig. 35).

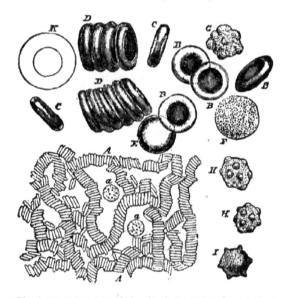

Fig. 35. Blood-corpuscles, magnified. At *A* the corpuscles are shown as seen when magnified four hundred diameters. The red corpuscles have adhered together by their flat sides, as they usually do soon after a drop of blood has been drawn. At *a* are seen colorless corpuscles. *B*, red corpuscles, very greatly magnified, seen in full face. *C*, the same seen edgewise. *D*, the same, adhering by their flat faces. *F*, *G*, *H*, colorless blood-corpuscles, very much magnified.

A few of the corpuscles are colorless and irregular in form (*F*, *G*, *H*, *I*), but by far the greater number are faintly colored. Seen by itself, each one looks pale yellow; but a number crowded together appear red.

6. What does a microscope show blood to consist of? Color and form of most of the corpuscles? Name? Number? Why is blood red? How may it be made yellow?

Hence they are called the *red corpuscles,* (*B, C, D*). In blood, the corpuscles are so closely packed that there are more than five millions in a single drop. It is this which makes the blood so red; if you dilute a drop of blood with a teaspoonful of water, or spread it out very thin on a piece of glass, it appears yellow.

7. The Shape of the Red Corpuscles is that of thin circular disks, a little hollowed out on each of their larger surfaces. If you made a piece of dough into a round cake, an inch across and a quarter of an inch thick, and then pressed it between thumb and finger so as to make a slight hollow on each side, you would have a very good model of a red blood-corpuscle. It would, however, be thirty-two hundred times broader and thicker than the real corpuscle. Put in another way, we may say that three thousand two hundred red corpuscles placed in a line, and touching one another by their edges, would make a row one inch in length; and twelve thousand eight hundred, piled one on another, would make a column an inch in height.

8. The Red Corpuscles of other Animals.—The red corpuscles of most mammalia (p. 9) resemble those of man

in being circular pale yellow disks slightly hollowed on each side; those of camels and dromedaries, however, are oval. The blood-

FIG. 36.—Red corpuscles of the frog.

corpuscles of dogs are so like those of man in size that

7. Shape of the red corpuscles? Illustrate. What is said of their size?

8. How do the red corpuscles of most mammalia resemble man's?

they cannot be readily distinguished; but in most cases the size is sufficiently different to enable a safe opinion to be formed, with a little pains. This fact has often been used to further the ends of justice, in determining whether spots of blood on the clothes of a suspected murderer were really due to the cause stated by him. The red blood-corpuscles of birds, reptiles, amphibians, and fishes, cannot be confounded with those of man, since they are oval and contain a little mass in the centre, which pushes out their sides and makes them project, instead of being hollowed.

9. The Use of the Red Corpuscles is to carry oxygen over the body. When blood flows through the lungs, these corpuscles take oxygen (Chap. XV.); as it flows through other organs they give up that gas to them. When the corpuscles are laden with oxygen their color is bright red, if a number of them be seen closely packed together; and when they have given up their oxygen, it is dark red. The different quantity of oxygen in the red corpuscles, is thus the reason of the different colors of arterial and venous blood.

10. The Blood-Plasma consists of water with a good many things dissolved in it. The most important of these are (1) albumens; (2) sugar; (3) minerals. The plasma has also floating in it many very small drops of

Exceptions? How may they be distinguished in most cases? How has this been used to further the ends of justice? Describe the red corpuscles of birds, etc

9. Use of the red corpuscles? When do they receive oxygen? When give it up? How does oxygen affect their color? Why do arterial and venous blood differ in color?

10. Of what does blood plasma consist? The most important things dissolved in it? Floating in it? What does the plasma contain in addition to nourishing substances?

fat. In addition to these nourishing substances, the blood which flows away from muscle or gland or brain contains some waste substances, which it is carrying off to the lungs or kidneys for removal from the body.

11. The Clotting or Coagulation of Blood.—When blood is first drawn from the living body it is perfectly liquid, flowing in any direction as easily as water. Very soon it becomes thick and sticky, like a red syrup; and at the end of five or six minutes it "sets" into a stiff jelly, which sticks to the cup or basin in which the blood is contained, so that the vessel may be turned upside down without spilling a drop. This alteration of the blood is named *clotting* or *coagulation*. It is due to a change of some of the dissolved albumens of the blood, into a solid substance named *fibrin*.

If the jelly be kept for half an hour or so, it shrinks and squeezes out a liquid named *serum*. The solid part floats in the serum and is named the *clot*.

12. The Use of Coagulation is to save us from the risk of bleeding to death from wounds. So long as the blood is flowing in healthy living blood-vessels, no fibrin forms in it, and it does not clot. But as soon as blood gets outside of the vessels, or whenever their lining is injured, clotting takes place. In this way, the ends of the small blood-vessels in a cut finger are soon clogged up, if we can only stop the flow for a little and give time for a clot to form in them.

11. What is the consistency of fresh blood? What changes occur in it during the first five or six minutes after it is drawn? What is the solidifying of the blood called? To what is it due? What is serum? What is the clot?

12. Use of coagulation? When does it not occur? When does it take place? Why does a cut finger stop bleeding after a short time?

13. The Heart (Fig. 37) resembles a pear in form, and is placed in a slanting position inside the chest, with its smaller end downwards. It lies just above the diaphragm (Fig. 2), and behind the lower two-thirds of the breast-bone. Its upper end, or *base* (so called because it is the

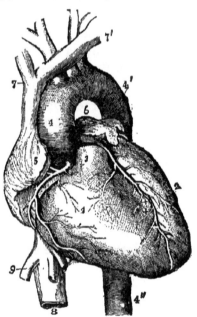

Fig. 37.—The heart and the arteries and veins opening into it, seen from the front. The pulmonary artery has been cut short close to its beginning. 1, right ventricle; 2, left ventricle; 3, root of the pulmonary artery; 4, 4', 4'', the aorta; 5, part of the right auricle; 6, part of the left auricle; 7, 7', innominate veins joining the upper vena cava; 8, inferior vena cava; 9, one of the veins from the liver, joining the inferior vena cava.

larger end, although the upper), projects a little to the right of that bone, and its lower end, or *apex*, a little to the left, where it may easily be felt beating by pressing with the finger between the cartilages (p. 18) of the

13. Shape and position of the heart? Where does its base project? Where may its apex be felt beating? Its size?

fifth and sixth ribs. A healthy heart is about the size of the clenched fist of its owner.

14. Interior of the Heart.—When the heart is cut open

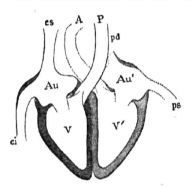

(Fig. 38) it is found to be hollow. Its cavity is not single, but is separated into a right and a left chamber, by a partition which runs through it from base to apex. Each chamber consists of an upper division, *Au*, *Au'*, called an *auricle ;* and a lower division, *V, V'*, named a *ventricle.* On each side there is a large opening between the auricle and ventricle; but

FIG. 38.—Diagram of a section through the heart. *Au, Au'*, auricles; *V, V'*, ventricles; *cs*, upper hollow vein; *ci*, lower hollow vein; *P*, pulmonary artery; *pd, ps*, pulmonary veins; *A*, aorta. Between each auricle and the corresponding ventricle, and at the mouth of pulmonary artery and aorta, the valves which control the direction of the blood-flow are seen.

there is no direct passage from the cavities on the right side of the heart to those on the left.

The divisions of the heart are, therefore, *right auricle, right ventricle, left auricle, left ventricle.*

15. The Pericardium.—The heart is surrounded by a loosely fitting case or bag, named the *pericardium.* The inside of this bag and the outside of the heart are covered by a very smooth membrane. In the space between the heart and its case there is in health a small quantity of

14 How is the cavity of the heart divided? Of what do its chambers consist? What is said of communication between auricles and ventricles of the same and of different sides? Name the divisions of the heart?

15. What is the pericardium? What is found inside it and outside the heart? In the space between? Use?

liquid, which makes the surfaces slippery, and allows the heart to contract or expand with very little friction.

16. The Vessels opening into the different Divisions of the Heart (Fig.38).—Veins bring back blood to the heart, and open into the auricles. Arteries carry blood from the heart, and start from the ventricles.

The veins pouring blood into the right auricle are named the *upper (cs)* and the *lower (ci) hollow veins*, or *venæ cavæ.* They got this name because after death, when most veins either collapse or are filled with blood, these are often found distended and empty.

One artery, *P*, springs from the right ventricle, and carries to the lungs the blood brought to the right side of the heart by the hollow veins. It is named the *pulmonary artery.*

The veins, *pd, ps*, which open into the left auricle are named the *pulmonary veins.* They bring blood from the lungs.

One artery, *A*, arises from the left ventricle; it is named the *aorta.* As it proceeds from the heart it divides, like the trunk of a tree, and at last its branches reach every organ of the body except the lungs.

17. The Beat of the Heart.—The heart relaxes about seventy times a minute, and takes blood from the veins; it contracts after each relaxation and forces blood into the arteries. This regularly alternating expansion and contraction, is known as the *beat of the heart.*

16. Function of veins? Place of opening into heart? Function and place of opening of arteries? What veins bring blood to right auricle? Why so named? What is the pulmonary artery? What veins open into left auricle? Whence do they bring blood? What is the aorta? What becomes of its branches?

17 Describe the beat of the heart?

18. The Valves of the Heart only permit blood to flow through it in one direction. When the heart is expanding and receiving blood, none flows back into the ventricles from the arteries, because the *semilunar (half-moon-shaped) valves*, at the mouths of the pulmonary artery and aorta, block the road. They will open outwards from the heart, and let blood *flow from* the ventricle, but they will not open the other way and let blood flow back from the artery into the heart.

When the heart dilates, it fills with blood from the veins. Then a ring of muscle round the mouth of each vein close to the heart, contracts and narrows the opening. Next the auricles contract, and the only way each can drive the blood collected in it, is into the ventricle of the same side. Immediately afterwards the ventricles contract and squeeze on the blood which has collected in them. This blood would go back into the auricles but for the valves which lie in the openings between each auricle and its ventricle, and only open towards the ventricle. As soon as any blood tries to flow back, the valves close and block the road, so the only way the contracting ventricle can send its blood is on into the arteries.

The valve between the right auricle and ventricle is named the *tricuspid* or *three-pointed valve*. That between the left auricle and ventricle is the *bicuspid* or *two-pointed valve*. It is sometimes named the *mitral valve*, from being shaped like the two points of a bishop's mitre.

18. Use of the valves of the heart ?   Position of semilunar valves ? Action ?   How are the mouths of the veins narrowed before the auri-cles contract ?   Where does each auricle pump its blood ?   What happens after the auricles have contracted ?   Why does not blood flow back in the auricles when the ventricles contract ?   Where do the ventricles pump blood ?   Where is the tricuspid valve ?   The mitral ?

19. The Course of the Blood-Flow. — Any portion of blood, starting from any chamber of the heart returns there after a short time, and starts from it again. This is why the blood-flow is called a *circulation*. The return is not direct; blood leaving the left side of the heart comes back first to the right, and blood starting from the right side returns first to the left.

How this occurs may be easily understood by examining Fig. 39, which represents, in a general way, the heart and blood-vessels. Starting from the *left* ventricle, *f*, blood flows along the aorta, *m*, and its branches, to all parts except the lungs.* These branches end in the very small and very numerous capillaries, *l*, of the muscles, and skin, and mucous membranes, and so forth. From these capillaries the blood is collected by veins, which unite to make the hollow veins, *k*, which pour it into the *right* auricle. From right auricle it is sent to right ventricle, and thence by the pulmonary artery and its

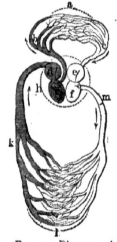

Fig. 39.—Diagram of the circulatory organs, to show the course of the blood-flow. *d*, right auricle of the heart; *g*, right ventricle; *b*, pulmonary artery and its branches; *a*, the capillaries of the lungs; *c*, the pulmonary veins; *e*, the left auricle of the heart; *f*, the left ventricle; *m*, the aorta, dividing into the smaller arteries; *l*, the capillaries of all the organs except the lungs; *k*, the veins bringing to the right auricle blood from all parts of the body but the lungs; *h*, the pericardium.

19. Why is the blood-flow called a circulation? To which side of the heart does blood which has last left the right ventricle first return? Starting from the left ventricle, describe the course taken by the blood until it gets back there. How often does the blood come back to the heart in making a complete circulation?

---

* Some branches of the aorta carry a little blood to the lungs; but for the purpose of getting a general idea of the blood-flow this may be neglected.

branches, *b*, to the lungs. There it flows through the pulmonary capillaries, *a*, and is collected from them into the pulmonary veins, *c*, which convey it to the *left* auricle; thence it flows to the left ventricle, and commences its round once more.

The valves of the heart only let the blood flow in the direction of the arrows in Fig. 39. If you start at any point in that figure and follow along in the direction pointed by the arrows, you will find that the blood cannot flow back at once, to the side of the heart it started from. To make a complete circulation, it twice leaves, and twice returns to, the heart.

20. The Systemic and Pulmonary Circulations.—To get from the left side of the heart to the right, the blood must flow through the arteries, capillaries, and veins of the body in general. This flow, from left ventricle to right auricle, is often named the *systemic circulation.* To get from the right side of the heart to the left, blood must flow through the arteries, capillaries, and veins of the lungs. This flow, from right ventricle to left auricle, is often named the *pulmonary circulation.* It is clear, however, that neither is a *circulation* in the proper meaning of the word, for after completing it, the blood is not back again at the place it started from. In order that it may be, it must go through both these circulations.

21. Illustration.—We may compare the blood-supply of the body to the water-supply of a city. The left side of the heart answers to the reservoir, and the arteries to

20. What is the systemic circulation ? The pulmonary ? Why is neither a circulation, strictly speaking ?
21 Compare the blood-supply of the body to the water-supply of a city. In what respects is it essentially different ? What would have

the water-mains. They begin at the heart, and are very much branched except close to it. The aorta answers to the main aqueduct leaving the reservoir, and there single, but giving off branches and becoming more and more divided the farther we follow it. At last the water-main ends in numerous but very much smaller tubes to supply various houses, as the branches of the aorta supply different organs.

The course of the blood differs, however, essentially from that of the water-supply of a city, for the used water does not return to the reservoir, whereas the blood is carried back to the heart. Instead of having a large supply of liquid stored up as in a reservoir, there is at any one time only quite a small amount in the heart, but this is steadily replaced by the inflow through the veins as fast as it is carried off by outflow through the arteries.

If the water used in the city were all carried back through the sewers (answering to the veins), to another reservoir placed beside the one it started from; and thence were carried by a different set of pipes (the pulmonary artery and its branches) into a purifying apparatus; and then back to the first reservoir, the whole process would be much like the circulation of the blood. The two reservoirs would represent the heart, which is double, and the purifying apparatus would represent the lungs.

22. The Pulse.—The arteries are as elastic as rubber tubing. Every time the heart beats and forces blood into them, their walls are stretched to make room for it. When an artery lies near the surface, this stretching

to be done with the used water to make the illustration complete? What would represent the lungs?

22. What is the pulse ?

can be felt through the skin. It is known as the *pulse.*
The number of pulses in a minute, therefore, tells the
rate at which the heart is beating.

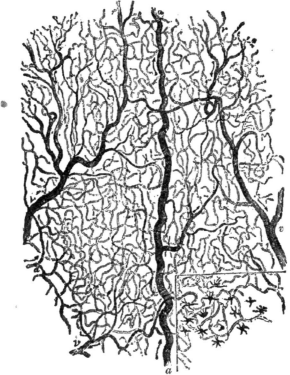

FIG. 40.—A small portion of the capillary network of the web between a frog's
toes, as seen with a microscope. *a*, a small artery feeding the capillaries; *v, v,*
small veins carrying blood back from the latter. The arrows indicate the direction of the blood-flow.

23. The Capillaries (Fig. 40) are such tiny tubes that
they cannot be seen without a microscope. Their vast
number makes up for their small size; in most organs

23. What is said of the size of the capillaries ? Of their number?
Illustrate. How does the blood do its work while flowing through
them ? Illustrate,

they are so closely packed that a pin's point cannot be
inserted without wounding some of them. This is illustrated when the skin is pricked. The blood in it is not
lying loose but is flowing in capillaries. We cannot in-
sert a needle deeper than the epidermis without wounding some of these capillaries and causing bleeding.

It is while flowing in the capillaries that the blood
does its work. Their walls are so thin that nourishing
matters easily soak through them to feed the organs;
and the waste matters of the organs readily pass
through the walls of these tiny vessels into the blood.

Imagine a piece of the finest net, with all its threads
consisting of hollow tubes, and diminished twenty
times in size, and you will have some idea of the
fine networks formed by the capillaries in the various
organs.

**24 Which Vessels contain Arterial and which Venous
Blood.**—As blood flows through the capillaries of the
lungs, its red corpuscles take up oxygen from the air
(Chap. XV.). The blood thus becomes bright red or arterial (p. 134). It flows, keeping this color, through the
left auricle and ventricle of the heart, and along the
aorta and its branches (the *systemic arteries*), which convey it to the body in general. These arteries pour the
blood into the capillaries of all organs except the lungs.
As it flows through these *systemic capillaries* the blood
gives up its oxygen to the organs and becomes dark-
colored. It is then collected ·into the *systemic veins*,
and, still of a dark color, is conveyed to the right auricle, right ventricle, and thence by the pulmonary artery

24. In what vessels does the blood become arterial ? Through
what part of its course does it keep its bright color ? Where does
it lose it ? Why ? Describe its course until it becomes bright

to the lungs, when it once more receives oxygen and becomes bright red.

Thus the pulmonary veins differ from all other veins in containing *arterial blood;* and the pulmonary artery and its branches, from all other arteries in containing *venous blood.* The ancient anatomists accordingly named the pulmonary artery, the *arterious vein.*

In Fig. 39, the vessels which convey venous blood are shaded.

again. How do the pulmonary veins differ from all other veins? The pulmonary artery and its branches from all other arteries?

APPENDIX TO CHAPTER XIII.

THE BLOOD.

Many of the main facts pertaining to the structure and composition of blood may be easily demonstrated as follows

1. Kill a frog with ether (p. 46); cut off its head, and collect on a piece of glass a drop of the blood which flows out. Spread out the drop so that it forms a thin layer. Hold the glass up against the light, and examine the blood with a hand-lens magnifying four or five diameters. The corpuscles will be readily seen floating in the plasma.

2. Wind tightly a piece of twine around the last joint of a finger, then, with a needle, prick the skin near the root of the nail. A large drop of blood will exude. Spread it out on a piece of glass and examine, as described above for frog's blood. The corpuscles will be seen floating in the blood-liquid, but not so easily as in frog's blood, since those of man are considerably smaller.

3. Obtaining a large drop of human blood as above described (2), note (*a*) that as it flows from the wound it is perfectly liquid ; (*b*) that it is red and very opaque; (*c*) spread it out very thin on the glass; note that it then looks yellow when held over a sheet of white paper; (*d*) mix a similar drop with a teaspoonful of water in a test tube; note that the mixture is yellowish, or, if not, becomes so on further dilution.

4. Place another large drop of human blood, obtained as above indicated, on a clean piece of glass   To prevent drying, cover by inverting over the drop a small glass whose interior has been moistened with water. In four or five minutes remove the wine-glass and note that the blood-drop has set into a firm jelly. Replace the moist

glass, and in half an hour examine again. The blood will then have separated into a tiny red clot, lying in nearly colorless serum.

5. If a slaughter house is accessible, the clotting of blood may be still better illustrated. Provide two large wide-necked glass bottles and a bundle of twigs. When the butcher bleeds an animal, collect in one bottle some blood, taking care that nothing else (contents of the stomach, for example, when the animal is bled, as is often done, by cutting its throat) gets mixed with it. Put this bottle aside until the blood clots, and carry it home with the least possible shaking. Next day the mass will exhibit a beautiful clot floating in serum. The latter will probably be tinted red, as the jolting in conveying the specimen from the slaughter-house shakes some of the red corpuscles out of the clot into the serum.

6. In the other bottle collect blood and stir it vigorously with the twigs for three or four minutes. Next day this specimen will not have clotted, but on the twigs will be found a quantity of stringy elastic material (fibrin), which becomes pure white when thoroughly washed with water.

7. Take some of the serum from specimen 5. Observe that it does not coagulate spontaneously. Heat it in a test-tube over a spirit-lamp, its albumen will then coagulate, like the white of a hard-boiled egg, and the whole will become solid.

8. Place a small quantity of whipped blood (6) on a piece of platinum foil. Heat over a spirit-lamp After the drop dries it blackens, showing that it contains much animal matter. As the heating is continued this is burnt away, and a white ash, consisting of the mineral constituents of the blood, is left.

The Circulatory Organs.

1. In the following directions "dorsal" means the side of the heart naturally turned towards the vertebral column, "ventral" the side next the breast bone, "right" and "left" refer to the proper right and left of the heart when in its natural position in the body; "anterior" means more towards the head in the natural position of the parts; and "posterior" the part turned away from the head.

2. Get your butcher to obtain for you a sheep's heart, not cut out of the bag (pericardium), and still connected with the lungs. Impress upon him that no hole must be punctured in the heart, such as is usually made when a slaughtered sheep is cut up for market.

3. Place the heart and lungs on their dorsal sides on a table in their natural relative positions, and with the windpipe directed away from you. Note the loose bag (*pericardium*) in which the heart lies, and the piece of midriff (*diaphragm*) which usually is found attached to its posterior end. .

4. Carefully dissecting away adherent fat, etc., trace the vessels below named until they enter the pericardium. Be very careful not to cut the veins, which, being thin, collapse when empty, and may be easily overlooked until injured. As each vein is found stuff it with raw cotton, which makes its dissection much easier.

a The *vena cava inferior* find it on the under (abdominal) side of the diaphragm; thence follow it until it enters the pericardium, about three inches further up; to follow it in this part of its course, turn the right lung towards your left and the heart towards your right.

The vein, just below the diaphragm, may be seen to receive several large vessels, the *hepatic veins.*

As it passes through the midriff, two veins from that organ enter it.

b. Superior vena cava seek its lower end, entering the pericardium about one inch above the entry of the inferior cava; thence trace it up to the point where it has been cut across; stuff and clean it.

c. Between the ends of the two venæ cavæ will be seen the two *right pulmonary veins,* proceeding from the lung and entering the pericardium; clean and stuff them.

5. Turn the right lung and the heart back into their natural positions; clear away the loose fat in front of the pericardium, and seek and clean the following vessels in the mass of tissue lying anterior to the heart, and on the ventral side of the windpipe.

a. The *aorta:* immediately on leaving the pericardium this vessel gives off a large branch; it then arches back and runs down behind the heart and lungs, giving off several branches on its way.

b. The *pulmonary artery:* this will be found imbedded in fat on the dorsal side of the aorta After a course, outside the pericardium, of about an inch, it ends by dividing into two large branches (right and left pulmonary arteries), which subdivide into smaller vessels as they enter the lungs.

c. Observe the thickness and firmness of the arterial walls as compared with those of the veins; they stand out without being stuffed

6 Notice, on the ventral side of the left pulmonary artery, the *left pulmonary veins* passing from the lung into the pericardium

7. Slit open the pericardiac bag, and note its smooth, moist, glistening inner surface, and the similar character of the outer surface of the heart. Cut away the pericardium carefully from the entrances of the various vessels which you have already traced to it. As this is done, you will notice that inside the pericardium the pulmonary artery lies on the ventral side of the aorta.

9. Note the general form of the heart—that of a cone with its apex turned towards the diaphragm. Very carefully dissect out the entry of the pulmonary veins into the heart. It will probably seem as if the right pulmonary veins and the inferior cava opened into the same portion of the organ, but it will be found subsequently (13, *a*) that such is not really the case. Note on the exterior of the organ the following points.

a. Its upper flabby *auricular* portion into which the veins open, and its thicker lower *ventricular* part.

b. Running around the top of the ventricles is a band of fat, an offshoot of which runs obliquely down the front of the heart, passing to the right of its apex, and indicating externally the position of the internal partition, or *septum,* which separates the right ventricle, which does not reach the apex of the heart, from the left, which does.

10. Dissect away very carefully the collection of fat around the origins of the great arterial trunks and that around the base of the ventricles.

11. Open the right ventricle by passing the blade of a scalpel through the heart about an inch from the upper border of the ventricle, and on the right of the band of fat marking externally the limits of the ventricles, and noted above (9, *b*), and then cut down towards the apex, keeping on the right of this line; cut off the pulmonary artery about an inch above its origin from the heart, and open the right auricle by cutting a bit out of its wall, to the left of the entrances of the venæ cavæ. On raising up by its point the wedge-shaped flap cut from the wall of the ventricle, the cavity of the latter will be exposed

a. Pass the handle of a scalpel from the ventricle into the auricle, and also from the ventricle into the pulmonary artery, and make out thoroughly the relations of these openings.

b. Slit open the right auricle. Observe the apertures of the *venæ cavæ*, and note that the pulmonary veins do not open into this auricle

12. Raise up by its apex the flap cut out of the ventricular wall, and if necessary prolong the cuts more towards the base of the ventricle until the divisions of the *tricuspid valve* come into view.

a. Note the muscular cord (not found in the human heart) stretching across this ventricle. Also the prolongation of the ventricular cavity towards the aperture of the pulmonary artery.

b. Cut away the right auricle, and examine carefully the *tricuspid valve*, composed of three membranous flexible flaps, thinning away towards their free edges; proceeding from near these edges are strong *tendinous cords* (*chordæ tendineæ*), which are attached at their other ends to muscular elevations (*papillary muscles*) of the wall of the ventricle

c. Slit up the right ventricle until the origin of the pulmonary artery comes into view. Looking carefully for the flaps of the semilunar valves, prolong your cut between two of them so as to open the bit of pulmonary artery still attached to the heart. Spread out the artery and examine the valves

d. Each flap makes, with the wall of the artery, a pouch, opposite which the arterial wall is slightly dilated. The free edge of the valve is turned from the heart, and has in its middle a little nodule (*corpus Arantii*).

13 Open the left ventricle in a manner similar to that employed for the right. Then open the left auricle by cutting a bit out of its wall above the appendage. Cut the aorta off about half an inch above its origin from the heart. The aperture between left auricle and left ventricle can now be examined; also the passage from the ventricle into the aorta, and the entry of the pulmonary veins into the auricle; and the *septum* between the auricles and that between the ventricles.

a. Pass the handle of a scalpel from the ventricle into the auricle; another from the ventricle into the aorta; and pass also probes into the points of entrance of the pulmonary veins. Observe that no other veins open into the left auricle.

b. Note the great thickness of the wall of the left ventricle, as compared with that of the right ventricle or of either of the auricles.

c. Carefully raise the wedge-shaped flap of the left ventricle, and
·cut on towards the base of the heart, until the valve (*mitral*) between
auricle and ventricle is brought into view; one of its two flaps will be
seen to lie between the auriculo-ventricular opening and the origin of
the aorta.

Examine in these flaps their texture, the chordæ tendineæ, the
papillary muscles, etc., as in the case of the right side of the heart
(12)

d. Examine the semilunar valves at the exit of the aorta; then cutting up carefully between two of them, examine the bit of aorta still
left attached to the heart, and note the valves more carefully as described in 12, *d.*

14. Examine a piece of aorta. Note that when empty it does not
collapse; the thickness of its wall; its extensibility in all directions;
its elasticity.

15. Compare with the artery the thin-walled flabby veins which
open into the heart.

CHAPTER XIV.

HYGIENE OF THE CIRCULATORY ORGANS.

1. To Ensure a Healthy and Regular Circulation of the blood, the skin must be kept warm. Cold, we have learned (p. 72), contracts its blood-vessels and drives the blood elsewhere. This does no harm if it be only for a short time; indeed often does good. But a prolonged chill of the surface is very apt to cause disease of some internal organ, by keeping it overfilled with blood, or *congested.*

A blush is a brief healthy congestion of the skin of the face. It may be compared to the flushing of the mucous membrane of the stomach when gastric juice is being secreted. In each case, the temporary rush of blood to the part nourishes it. But continued overfulness of blood has an opposite effect. Too much liquid from the blood soaks through the walls of the capillaries, and the organ in which they lie becomes puffy and swollen.

2. Taking Cold.—Congestion produced by a chill of the surface oftenest shows itself on some mucous membrane. If that lining the nose be attacked, it becomes

1. What effect has the temperature of the skin on the circulation of the blood? What is apt to result from a prolonged chill of the surface? What is a blush? To what compared? Results of prolonged overfulness of blood?
2. Where do congestions due to cold most often occur? Describe

swollen, and we have difficulty in breathing through the
nostrils.   It is also irritated, and so we sneeze (p. 174).
Unless proper means be at once taken to stop the "cold,"
the congested mucous membrane becomes *inflamed*.   In
that case, its vessels are not only gorged with blood, but
the whole membrane is in a state of unhealthy activity.
So far as its glands are concerned, this is shown by the
unnaturally abundant watery mucus which runs from
the nostrils.

When deeper parts of the mucous membrane are
attacked by "a cold," we cannot observe the details so
easily.   But they are much the same in all cases.   Thus
when the mucous membrane of the tubes which carry air
into the lungs (p. 170) is the one attacked, we suffer from
a "cold on the chest," or *bronchitis*.   In this case we
have difficulty in breathing, because the swollen membrane narrows the air-passages; we feel pain and irritation in the chest; and we cough up abundant "phlegm"
or unnatural secretion.

If the "cold" attacks the mucous membrane of the
stomach, we suffer from loss of appetite and from indigestion, because the altered secretion fails to do its
proper work.   The production of diarrhœa by cold attacking the bowels has been already referred to (p. 129).

3. To Avoid taking Cold, the essential things are to
wear proper clothing, and, when perspiring, to guard
against sudden cooling (Chap. VII.).   If unavoidably
exposed to conditions apt to cause a cold, the risk may

the condition of the mucous membrane of the nose during a " cold in
the head."   That of the air-passages during a " cold on the chest."
Results of a cold attacking the mucous membrane of the stomach?
Of the bowels?

3. To avoid taking cold what things are most necessary?   What
should be done to prevent a cold, after exposure likely to cause one?

be much diminished by prudence.   Try to get your skin warm and your sweat-glands active as soon as possible. Exercise is usually the best way to do this.   When you feel chilled, and have some distance to go before you can reach a warm room or get extra clothing, it is wiser to run or walk or row, if possible, than to sit still and be driven.   The muscular exercise will warm the skin.   If obliged to keep on wet clothing, throw over it a dry wrap.   This will prevent the wet garments from drying rapidly, and thus taking heat from the skin too fast (p. 67).   As soon as possible rub the whole skin briskly until it is red and warm; then put on dry woollen clothing.   If your skin does not quickly warm when rubbed, take a warm bath, go to bed, and drink two or three large cups of hot weak tea or lemonade.   If a warm bath cannot be had, put the feet in hot water.

4. **Articles of Dress should not Fit so Tightly as to Check the Blood-Flow.**—Most large arteries lie deep, but many large veins are near the surface, just under the skin. The flow of blood in a vein is easily stopped by pressure, because the walls of the veins are thin and flabby; and when the vein leading from any organ is squeezed, the blood-flow from it is hindered.   Thus congestion is produced.

The veins most often impeded in their work by tight clothing, are those of the neck and leg.

The chief veins bringing back blood from the head are the *external jugular veins*, which lie under the skin,

What if the clothing is wet and cannot be at once changed ?   What should be done as soon as you can change it ?   What is the object of the exercise, baths, rubbing, etc.?

4. How may tight garments produce congestion ?   Which veins are most often compressed by articles of ordinary clothing ?   What are the external jugular veins ?   What is apt to follow if they are com-

one on each side of the neck. A tight collar or scarf compresses these veins and tends to cause congestion of the brain, dizziness, redness of the eyes, and a flushed face.

The chief vein which brings back blood from the foot and the lower leg is named the *long saphenous vein* (Pl. IV., 28). It begins on the inner side of the ankle, and runs to the top of the thigh. A tight garter compresses the saphenous vein, into which many other veins of the leg pour their blood, and thus checks the circulation in that part of the body. The results are deficient blood-flow in the feet. Congestions and inflammations, as *chilblains*, more easily occur, and the feet are more apt to become cold. If the garter be very tight, the veins below it often get so gorged with blood that their walls stretch and form swellings, known as *varicose veins.* Varicose veins sometimes burst and cause dangerous bleeding; they very often so press and crush the tissues in their neighborhood as to cause inflammation and sores. The stocking-supporters now so commonly used, which attach the stocking to the waistband, are far better than ordinary garters.

5. Muscular Exercise Promotes the Circulation of the Blood, not only because it quickens the beat of the heart, but because the contracting muscles drive along the blood in the veins.

In the veins are numerous valves (Fig. 41), which open towards the heart and from the capillaries. Blood flow-

pressed? What is the saphenous vein? Describe its course. How does a tight garter affect the flow of blood in the leg? Results? How may varicose veins be produced? Consequences of varicose veins?

5 How does muscular exercise promote the blood-flow? How is blood prevented from flowing back through the veins towards the

ing in the proper direction, *A*, from capillaries to heart, is not hindered on its road; but any back-flow in the opposite direction, *B*, is at once checked by the closing of the valve. This you may easily observe on the back of the hand of any one who is thin. Select a vein which has no branches for an inch or more. Press on its lower end, that is the end nearer the fingers, so as to close

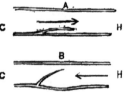

Fig. 41.—Diagram to illustrate the mode of action of the valves of the veins. *C*, the capillary, *H*, the heart end of the vessel.

it. Then push the blood out of it by rubbing it with your forefinger in a direction towards the wrist. The vein will remain empty up to the place where the next higher branch joins it. At that place there is a valve, which will not allow blood to flow back into it. As soon as you remove the pressure from its lower end, however, the vein immediately fills, with blood brought to it from the capillaries of the fingers. We learn from this simple experiment that the valves of the veins allow blood to flow through the veins to the heart, but *not from* it. If the first vein you try the experiment with, does not behave as it should, try another, for some of the veins on the back of the hand have branches entering their deeper sides, which you cannot see, and from which they become filled.

When the muscles contract in length and swell in breadth (p. 35) they press on the veins near them. This pressure cannot drive blood back to the capillaries, on account of the valves in the veins. But it drives blood on from the veins towards the heart, and thus pro-

capillaries? How may we observe on the hand the action of the valves of the veins? How do the muscles promote the circulation of

motes the circulation. When the muscles relax, the veins fill again; and then the next muscular contraction forces the blood inside them on towards the heart. In this way muscular exercise is a great· help to the heart in keeping up the flow of blood.

When you feel cold, a brisk walk or run, or, if the weather is too severe for outdoor exercise, indoor gymnastics, will warm you sooner and better than sitting over a fire. This is especially the case with coldness of the feet. Toasting them over a fire is of little use. They become cold again almost as soon as you leave the fire. But a brisk walk or an active game will soon increase the circulation, and make the feet warm for the rest of the day.

6. Cuts and Wounds.—If the wound be made by a clean sharp instrument and the bleeding is not great, press its edges together and hold them in place by a moderately tight bandage. The edges of a gaping wound may need to be held together by sticking-plaster; in other cases it does no good. Wounds which a single wider strip of plaster will not hold at all, may be nicely held together by separate narrow strips from $\frac{1}{4}$ to $\frac{3}{8}$ of an inch wide, according to the nature of the cut. Taking one, warm and fasten it on the farther side of the wound. Pull the loose end of the strip towards you and press the nearer lip of the wound against the farther one, then fasten the rest of the strip firmly down and hold it till it sets. Proceed in like way with the other strips. Ointments and salves are never necessary to promote the healing of a simple clean cut, and very often do harm.

the blood? How does this affect the heart? Why is it better to warm yourself by exercise than by sitting over a fire?
6. What is the proper treatment for a "clean" cut? If its edges gape? How should sticking-plaster be put on large cuts? What is said of ointments and salves? What should be done if there is dirt in

If the cut has been made by a sharp instrument but has dirt or grit in it, hold its edges apart and wash by pouring water on it. Then proceed as above. Do not sponge or wipe it. Either cold water or water as hot as the hand can bear may be used. Both check bleeding, the hot water rather better than the cold. Tepid water promotes bleeding.

A jagged cut, or a wound made by a blunt instrument, does not heal as easily as one made by a sharp knife. If it is large, or is on a part of the body where it is very desirable to avoid a scar, send for a doctor. Meanwhile, if blood oozes out fast, check the bleeding by constant pressure with sponges wrung out of hot water.

7. **Wounds of Large Arteries or Veins** need prompt treatment, lest the sufferer die from loss of blood. If a big vein has been divided, the blood will flow out pretty steadily and of a dark color. If a large artery has been cut, the blood will be brighter red and probably come out in spurts. Whichever it may be, the proper thing is to send at once for medical aid, and, until it arrives, to stop the bleeding. Do not lose time by trying to decide whether the flow is from a vein or an artery, and whether you should apply pressure nearer or farther from the heart. Many large arteries and veins when cut bleed nearly as fast from one end as the other. Press at once on the wound as hard as you can, with a handkerchief or anything of the sort at hand; and, when you have thus partly checked the bleeding, try pressure all around the cut,

the wound? If the cut is jagged or apt to be disfiguring? Until the doctor arrives?

7 Why should wounds of large blood-vessels be treated at once? How does the flow from a vein usually differ from the bleeding of a wounded artery? What had best be done in either case? How proceed until skilled advice is obtained? How may the blood-flow from

above and below, and on each side, till you find the place where it "does most good." In deep wounds of the arms or legs, you will usually find that pressure both above and below is necessary. A surgeon would know where to apply the pressure in the case of any particular wounds, but you do not: your business is to find it out by experiment as soon as you can, and not trouble yourself with any general rules, which will fail you in most particular instances.

If the wound is on the lower part of a limb and you find that you cannot entirely check the loss of blood by pressure on it and in its neighborhood, keep up the pressure and get some one to bind the limb very tightly higher up. This is best done as follows: Tie a handkerchief loosely round the upper part of the wounded arm or leg; then put a stick under it, and twist the stick round and round until the handkerchief is so tight as to close the arteries, and stop all flow of blood to the lower parts of the limb. Such stoppage of the blood-flow for half an hour or even a little longer, will do no permanent harm, while free bleeding from a wound in a large artery or vein may cause death in three or four minutes.

If a person who has lost much blood begins to breathe slowly and irregularly, give him a strong stimulant as soon as you can get it, and choose the stimulant you can get quickest. If a drugstore is close by, a mixture of a teaspoonful of aromatic spirits of ammonia with table spoonful of water may be given. If brandy or whiskey can be obtained sooner, use them. The irregular breathing is a sign that the part of the nervous sys-

a wound in the lower part of the arm or leg be stopped? If the sufferer shows signs of death from loss of blood, what should be done? What is the use of the stimulant in this case?

tem (Chap. XVIII.) which makes the muscles of breathing do their work, is ceasing to act, and extra stimulation must be given it for a while until the bleeding is stanched and blood has again commenced to collect in the arteries.

8. The Action of Alcoholic Drinks on the Circulation.— Alcohol in excess, injures the blood, the arteries, and the heart.

Even in moderate doses, it diminishes the power of the blood to absorb oxygen, and thus decreases the oxidations within the body, and lowers its working power and its temperature. Large quantities of alcohol cause the red blood-corpuscles to become shrunken and distorted, and greatly diminish their efficiency as carriers of oxygen to all the organs.

9. The Action of Alcohol on the Plasma.—Continued alcoholic indulgence leads to an alteration in the blood-plasma, lessening its tendency to form fibrin and to clot. Hence even the slight wounds of tipplers are apt to result in dangerous bleeding The fibrin is so scanty and the clogging up of the ends of cut blood-vessels so slow (p. 138), that all surgical operations on such persons are attended with special danger.

10. The Action of Alcohol on the Arteries.—Alcohol tends to make fatty matter collect in the walls of the arteries. The oil-drops take the place of the natural tough elastic material. Thus the artery is weakened. In consequence,

8. Action on the blood of even moderate doses of alcohol? How does this affect the body? Action on the red corpuscles of large doses of alcohol?

9. Action on the blood-plasma? Why are surgical operations on tipplers especially dangerous?

10. Action of alcohol on the arteries? What is an aneurism? Usual result?

the blood, which is forcibly sent into it by the heart, may stretch its walls and make it swell out and become thin. Such a swelling on an artery is named an *aneurism.* An aneurism usually ends by bursting, and the person bleeds to death.

11. The Action of most Alcoholic Drinks on the Heart is to excite it and hurry its beat. Whether pure alcohol, diluted with water, has this action, is not certain. It is certain that most ordinary alcoholic drinks, as wines and spirits, have it. When the beating of the heart is quickened, each contraction of its muscles takes about as long as when it beats slower, but the time of repose between the beats is shortened. The result is that the heart is overworked. It has not sufficient rest for its proper nourishment, and gradually undergoes a change known as *fatty degeneration*   Fatty or oily matter takes the place of the proper muscle-substance, and the heart, becoming more and more weakened, at last cannot pump the blood over the body. The consequence, of course, is death. Fatty degeneration of the heart is so often due to indulgence in alcoholic stimulants, that a fatty heart is often called by physicians a "whiskey-heart.'

11.   What is the action on the heart of all ordinary alcoholic drinks? How is the resting time of the heart affected when its beat is quick-ened?' Result? What is fatty degeneration? Consequence when it occurs in the heart?

CHAPTER XV.

RESPIRATION OR BREATHING.

1. The Use of Respiration is to renew the air in the lungs. This is necessary because the blood, as it flows through the lungs, is all the time taking something from the air within them, and giving something to it. If this air were not passed out, and fresh air taken in its stead, it would soon have nothing left of what the blood wants. It would also become so loaded with the waste matter the blood gives off to it, that it could take no more of it, and so the blood would not be purified. *Suffocation* is death from want of fresh air in the lungs.

2. What the Blood Takes from the Air.—Blood gets nourishment for the body from the alimentary canal. But we have learned (Chap. VIII.) that in order that foods may give us power and keep us warm, they must be oxidized, and, clearly, they cannot be oxidized unless oxygen is supplied. The blood, into which the digested foods are taken, as it flows through the lungs absorbs this necessary oxygen.

3. What the Blood Gives to the Air.—The blood in its passage through the lungs gives off to the air, heat,

1. What is the use of respiration? Why is it necessary? What is suffocation?
2. Where does blood get nourishment for the body? Why is oxygen necessary as well as food? Where does the blood get oxygen?
3. What does blood give to the air as it passes through the lungs?

water, a gas named *carbonic acid,* and a small quantity of *organic matters.*

The heat is easily recognized : you know that your breath is warm. The water usually comes out in the form of invisible vapor. On a cold day, however, it is seen as mist, streaming from the nostrils ; and any day it can be made visible by breathing on a cold bright object, as a mirror or knife-blade.

The carbonic-acid gas and the organic matters given out in the breath, are unfortunately not so easily made apparent as the heat and the water-vapor. They are, however, of very great importance. Carbonic acid is one of the chief waste substances made by the body and must be removed from it. The organic matters poison the body, if air containing them be breathed over and over again.

4. How Carbonic Acid is made from Charcoal.—You remember that when a human body is incompletely burned (Chap. I.) it forms a black mass of charcoal. Now *charcoal* is a mixture of a substance named *carbon* with some minerals. We may call it impure carbon. When it is burned, its carbon combines with oxygen and makes carbonic-acid gas. Just as rust is oxidized iron, so carbonic acid is oxidized carbon ; though it is a gas, instead of being solid like iron-rust.

5. How Carbonic Acid is made in our Bodies.—All our organs contain animal matter (p. 10), and all animal

How may the heat be recognized ? How the water ? Name a chief waste matter produced by the body. What is the result of frequently breathing air containing organic matters given out in the breath ?

4. What is left when a human body is incompletely burned ? What is charcoal ? What may we call it ? What happens when it is burned ? How does carbonic acid resemble and differ from iron-rust ?

5. What do all our organs contain ? What do all animal matters

matters leave charcoal when they are partly burned. They must, therefore, contain carbon. It seems odd that this should be so, and yet that they should not be black; but carbon is not always black. A diamond is nearly pure carbon; and can, by being heated, be changed into ordinary black carbon, and then burnt and combined with oxygen to make carbonic acid. Very few substances which contain carbon combined with other things, are black; as, for example, carbonic-acid gas itself, which is quite colorless. In our organs, the carbon is all combined with other things; it only shows its black color when the heat of the fire has separated them from it.

As long as we live, our bodies are slowly burning or oxidizing (Chap. VIII.). By this burning, carbonic acid is produced from the carbon of their organs. When we work hard, a great deal is made, and when we are at rest much less. But even in 'deep sleep, oxidation is going on inside our bodies all the time, to supply animal heat, and force or power for every heart-beat, and each movement of breathing. The carbonic acid produced by oxidation in the body, must be removed. When it is abundant, the organs cannot receive or use the oxygen which they need in order to do their work.

6. Breathing Air which contains much Carbonic Acid, will not Support Life.—The more carbonic acid in the air, the less oxygen. If there is very much carbonic

leave when they are partly burned? What must they, then, contain? Illustrate the fact that things containing carbon are not always black. In what state does carbon exist in our bodies? When does it show its black color? What occurs as long as we live? What is produced by the burning? Why is oxidation necessary even during deep sleep? Why must the carbonic acid produced in the body be removed?

6. How does the presence of carbonic acid in air affect the quantity of oxygen? What is the consequence to life if there is much of

acid, there is not enough oxygen to supply the needs of the body and maintain life. Death in such case results from suffocation, which may be more plainly named *oxygen-starvation.* The "foul air" which is sometimes present at the bottom of deep wells or pits, and kills people who incautiously go down them, does so because it contains much carbonic acid. Carbonic acid is not itself very poisonous, but air containing much of it is fatal, because carbonic acid has taken the place of the necessary oxygen; and air without plenty of oxygen will not support life.

7. Excretion.—The process of removing its wastes from the body, or getting rid of things which have done their work in it and are no longer wanted, is named *excretion.* The waste substances themselves are also called *excretions.* Organs which remove them are *excretory organs.*

8. The Lungs Perform a Double Duty.—So far as oxygen is concerned, they are organs for taking something useful into the body, and are *receptive organs.* So far as carbonic acid is concerned, they are *excretory organs.*

9. How the Air is Purified.—Every living human being and every one of the lower animals is all the time taking oxygen from the air and giving carbonic acid to it. So is every fire and every burning candle. We may natutally ask, How is the air kept fit to breathe?

it? Give another name for suffocation? Why may carbonic acid be called a "negative poison"?

7. What is the process of excretion? What substances are named excretions? What are excretory organs?

8. What are the two main duties of the lungs?

9. How do living animals and fires alter the air? What living things purify the air? Name a chief food of green plants. What do they do with the carbonic acid they take from the air? How do animals and plants help one another?

The air is purified by plants. All green plants, when in the light, take up carbonic acid from the air. It is one of their chief foods. From the carbonic acid, they pick out the carbon and use it in making starch and sugar and oils, and other things. The oxygen, they give back to the air. Thus plants not only make food for animals but keep the air fit for them to breathe; while animals by their breathing supply food for plants.

10. The Air inside the Lungs must be Frequently Changed.—If the air inside the lungs be not frequently replaced by fresh air, it becomes so full of carbonic acid that it can take no more from the blood; and so poor in oxygen that it cannot supply the blood with enough of that gas. Dark-colored venous blood comes to the lungs by the pulmonary artery (Chap. XIII.), containing little oxygen and much carbonic acid. Through the thin walls of the pulmonary capillaries, it gives carbonic acid to, and takes oxygen from, the air inside the lungs, and thus, replenished and purified, is returned to the left side of the heart to be distributed over the body.

11. The Respiratory Organs are—(1) the *lungs*, in which the blood is exposed to the action of the air; (2) the *air-passages*, through which air enters and leaves the lungs; (3) certain muscles (*muscles of respiration*), and the skeleton of the thorax, which work together to alternately expand and contract the chest, and thus renew the air inside the lungs.

10. What happens if the air within the lungs is not frequently renewed? What sort of blood does the pulmonary artery bring to the lungs? What does this blood do as it flows through the pulmonary capillaries? What becomes of it after leaving the lungs?

11. Name the respiratory organs. Use of the lungs? Of the air-passages? Of the respiratory muscles and the skeleton of the chest?

12. The Air-Passages are the *nostril-chambers*, the *pharynx*, the *larynx*, the *windpipe* or *trachea*, the *bronchi*, and the *bronchial tubes*. The nostrils (Fig. 42) open behind, into the upper part of the pharynx. From the front of the pharynx, below the level of the root of the tongue, *k*, and above the opening into the gullet, the *larynx* proceeds. Its opening is overhung by a sort of lid, *e*, named the *epiglottis*. This lid shuts down when food or drink is passing through the pharynx, but stands up at other times.

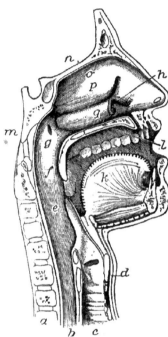

FIG. 42.—The mouth, nose, and pharynx, with the commencement of the gullet and larynx, as exposed by a section, a little to the left of the middle of the head. *a*, vertebral column; *b*, gullet; *c*, windpipe; *e*, epiglottis; *f*, soft palate; *g*, opening of Eustachian tube; *k*, tongue; *l*, hard palate; *m*, the sphenoid bone on the base of the skull; *n*, the fore part of the skull-cavity; *o, p, q*, the turbinate bones of the outer side of the left nostril-chamber.

13. The Larynx is the organ in which *voice* is produced. The hard projection in front of the neck, commonly named "Adam's apple," is caused by the larynx, which is a cartilaginous box (*d*, Fig. 42) lined by mucous membrane. At one place the mucous membrane is pushed in from each side, so that only a narrow slit is left in the middle. This

12. Name the air-passages? Into what do the nostrils open behind? Where does the larynx begin? What is the epiglottis? Its use?

13. Name the organ of voice? What is the larynx? What is the glottis? What are the vocal cords?

slit (*c*, Fig. 43) is named the *glottis*. The folds forming
its sides are elastic and tightly stretched ; they are
named the *vocal cords*.

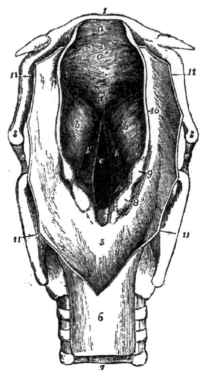

FIG. 43.—The interior of the larynx as seen when viewed from above and behind,
through its opening into the pharynx. 1, hyoid bone, which in life has the root of
the tongue attached to it ; 5, lower part of the pharynx cut open ; 6, top of the
gullet ; 8, 9, 10, the right edge of the opening of the larynx ; *a*, *a'*, *a''*, epiglottis ;
c, glottis (the dotted lines leading from the letter point to the edges of the vocal
cords) ; *b'*, *b'*, hollows in the mucous membrane of the larynx, above the vocal
cords ; *b*, *b*, rounded prominences of the mucous membrane, named the false
vocal cords : they play no direct part in the production of voice.

14. How Voice is Produced.—Certain muscles separate
the vocal cords and widen the glottis ; others bring the

14. How do muscles alter the glottis ? What is its state in quiet

cords together and narrow the glottis.   In ordinary quiet breathing the glottis is wide open, and air passes through it without causing sound.   When it is narrowed, and air driven through it from the lungs, voice is produced.

The sounds produced in the larynx are afterwards altered, and added to, in various ways in the throat, mouth, and nose.   Thus voice is altered or improved into *speech*.

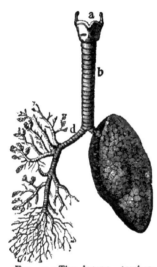

15. The Windpipe or Trachea (*b*, Fig. 44) is a stiff tube which may be easily felt in the lower part of the front of the neck of thin persons.   In its walls are horseshoe - shaped cartilages, which keep it open.   The windpipe enters the thoracic cavity, and there divides into two *bronchi*, one for each lung (*d*, Fig. 44).

FIG. 44.—The larynx, trachea, bronchi, and bronchial tubes, seen from the front.   The right lung (to the left in the figure) has been dissected away to expose the bronchial tubes.   *a*, larynx; *b*, windpipe; *d*, right bronchus: its branches are bronchial tubes.

16. The Bronchial Tubes. — Each bronchus, as soon as it enters the lung, begins to divide, over and over again, like the trunk of a tree.   The branches are hollow, and the end ones are very small indeed.   They are all named *bronchial tubes*.   On the left side of Fig. 44 the right lung has been cut away, so as to show the bronchial tubes.

breathing?   How is voice produced?   How is voice converted into speech?

15.   What is the windpipe?   How is it kept open?   Where and how does it end?

16.   What is said concerning the bronchial tubes?

17. The Lungs lie inside the thorax, one on each side of the heart (Fig. 2). They are elastic spongy masses, full of tiny cavities, named *air-cells*. Into the air-cells the smallest bronchial tubes open (Fig. 45). Thus air gets to them, ready to give oxygen to the blood, and carry off carbonic acid from it.

FIG. 45.—A small bronchial tube, *a*; *b*, *c*, air-cells connected with it. Magnified about twenty times.

18. Inspiration and Expiration. — Breathing consists of breathing-in and breathing-out, turn and turn about. Breathing-out gets rid of air which has become foul in the lungs: it is named *expiration*. Breathing-in conveys new air to the lungs in place of that which has been expired: it is known as *inspiration*.

19. The Movements of the Chest alternately enlarge and diminish its cavity. When it is enlarged, air enters it; when it is diminished, air is driven out. We may compare the chest in this respect to a pair of bellows. The chief difference is that air enters the bellows through one aperture, and is driven out through another; while in breathing, air comes and goes by the same road, the windpipe, which answers to the nozzle of the bellows.

20. How the Chest-Cavity is Enlarged to cause Inspiration.—The enlargement of the chest is brought about

17. Position of the lungs? Structure? How does air reach the air-cells of the lungs?

18. Of what does breathing consist? Use of breathing-out? Its technical name? Of breathing-in? Its technical name?

19. Result of the chest-movements? What happens when the chest is enlarged? Diminished? Illustrate. What part of the respiratory organs corresponds to the nozzle of a pair of bellows?

20. How is the chest enlarged? Position of the ribs in expiration?

by certain muscles which move the ribs, and by the diaphragm.

The ribs during expiration slope downwards, the end of each attached to the spinal column being higher than the end attached to the breast-bone (Fig. 5). When we draw a breath, certain muscles pull up the front ends of the ribs. When this occurs, the breast-bone is pushed farther away from the back-bone, and the depth of the chest between breast-bone and spinal column is increased.

That raising the front end of the ribs must push the breast-bone forwards may be readily understood by examining Fig. 46. In the figure, *ab* represents the spinal column, and *st* the breast-bone. The position of the ribs in expiration is indicated by the rods *c* and *d*; their position in inspiration, by the dotted lines *c'* and *d'*. It is clear that when the ribs are raised the sternum must be separated farther from the back-bone.

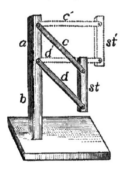

Fɪɢ. 46.—Diagram of a model to illustrate how the chest-cavity is increased from before back when the front ends of the ribs are raised. *ab* represents the spinal column; *st*, the breast-bone; *c, d*, the ribs in expiration; *c', d'*, the ribs in inspiration.

21. Action of the Diaphragm during Inspiration. — The diaphragm (*d*, Fig. 1) is a dome-shaped muscle, with its hollow side turned towards the abdomen. When it contracts, it flattens, and thus increases the chest-cavity. At the same time, it pushes down the liver, stomach, and intestines. These make

In inspiration? What is the result of raising the front ends of the ribs?

21. Form of the diaphragm? How altered when it contracts? Result as regards the chest? The abdomen?

room for themselves by pushing out the soft front wall
of the abdomen, which therefore protrudes.

**22. The Combined Action of the Diaphragm and of the
Muscles which Raise the Ribs** is such as to considerably
increase the chest-cavity. This is illustrated in Fig. 47.
In *B* are shown the size and form of the thoracic
cavity, and the position of the diaphragm, after an expi-

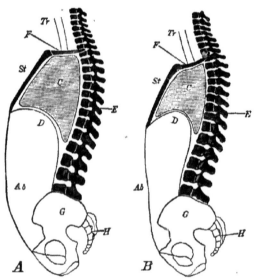

FIG. 47.—Diagrams showing the form and size of the chest and abdomen during
inspiration, *A*, and expiration, *B*. *C*, chest-cavity; *D*, diaphragm; *E*, spinal column; *F*, collar-bone; *St*, sternum; *Ab*, abdomen; *G*, hip-bone; *H*, coccyx.

ration. *A* represents the chest and diaphragm at the
end of an inspiration.

23. Expiration.—In expiration, the chest-cavity is diminished, and air driven out of the lungs. It is ordinarily brought about without muscular work. The mus-

22. What is the result of the combined action of the diaphragm and
the muscles raising the ribs during inspiration?

23. What happens during expiration? How is it ordinarily brought
about? Explain?

cles which have pulled up the ribs and sternum (to cause inspiration) relax, and these bones fall back into their former places. The diaphragm also relaxes, and the liver, stomach, and intestines, pressing against its underside, then push it up towards the chest. Thus the lungs are squeezed and air driven out of them.

24. Sneezing and Coughing.—The mucous membrane lining the nose and the larynx is very sensitive. Anything irritating· it causes a peculiar kind of violent expiration, calculated to drive a powerful blast of air through the air-passages and force away the irritant. When the inside of the nose is tickled, a *sneeze* follows. The irritation makes us first draw a deep breath, without our willing it at all, and in spite of our will if we try to prevent it. Then, when the lungs are filled with air, the glottis (p. 169) is closed and the chest compressed. Next, the glottis is suddenly opened and the compressed air rushes out of the lungs. It is made to go through the nose, because the root of the tongue and the soft palate are brought together, so as to close the opening from the pharynx to the mouth. Sneezing is a good example of the resemblance of our bodies in many ways to machines, made to do a certain thing under certain circumstances. The control which we have over them by our will is not at all complete. We can neither prevent a sneeze when the nose is irritated, nor make even a good imitation of a real sneeze when it is not.

A *cough* differs from a sneeze, mainly in the fact that the air is allowed to pass out through the mouth. Its use is to drive out anything irritating the larynx.

24. What results when the mucous membrane of nose or larynx is irritated? Describe the process of sneezing? What may we learn from it? Of coughing? Its use?

·APPENDIX TO CHAPTER XV.

1. A sheep's lungs with the windpipe attached may be readily obtained from a butcher. It is best to secure them and the heart all together, as unless the heart be carefully removed holes are apt to be cut in the lungs.

2. Examine the windpipe, and trace it down to its division into the bronchi. In the wall of the windpipe note the horseshoe-shaped cartilages which keep it open, and which are so arranged that the dorsal aspect of the tube (which lies against the gullet) has no hard parts in it.

3 Trace the main right bronchus to its lung, and then, cutting away the lung-tissues, follow the branching bronchial tubes through the organ. Note the cartilages in their walls In the sheep there is a small extra bronchus on the right side, which goes to the upper part of the right lung. It is not present in man.

4. Carefully divide the left bronchus where it joins the windpipe, and lay it and its lung aside. Then slit open the trachea, the bronchus still attached to it, and the bronchial tubes. Observe the soft pale-red mucous membrane lining them.

5. In the left bronchus, which has still an uninjured lung attached to it, tie air-tight a few inches of glass tubing of convenient size. On the end of the glass tube then slip a few inches of rubber tubing On blowing through the rubber tube the lung will be distended, and as soon as the opening is left free it will collapse; in this way its great extensibility and elasticity will be seen.

6. Blow up the lung moderately, and while it is distended tie a string very tightly around the bit of rubber tubing. This will keep the air from escaping; the distended lung can now be examined at leisure, and its form, lobes, and the smooth moist pleura covering it be better seen than when it is collapsed.

7. The diaphragm may be readily seen in the body of any small animal (rat, kitten, puppy), on removing the abdominal viscera. The liver and stomach must be cut away with especial care.

a. When the above viscera are removed, the vaulted diaphragm will be seen, and through it the pink lungs.

b. Pull the diaphragm down, imitating its contraction and flattening in inspiration. The lungs will be seen to follow it closely, expanding to fill the space left by it in its descent.

c. Make a free opening into one side of the thorax. The corresponding lung will collapse, and be no longer influenced by movements of the diaphragm.

d. Now open the other side of the chest· its lung also shrinks up; the structure of the diaphragm (its tendinous centre and muscular sides) can now be better seen, as also the attachment of the pericardium to its thoracic side.

CHAPTER XVI.

HYGIENE OF RESPIRATION.

1. Introductory.—We all know, of course, that air which is not fresh is unpleasant to breathe, but many persons appear not to know that it is also poisonous.

Suppose you put an air-tight bag, containing two or three pints of air, close to your mouth, and kept your nostrils closed, so that no air could enter the lungs but that in the bag. For the first few breaths you would have no trouble. But after you had breathed in and out of the bag several times the air within it would not have enough oxygen left to supply the needs of the body, and would be so full of excretions as to be poisonous.

If you want to keep a pet puppy or kitten in a box, you make an air-hole. When asked why, you reply that the animal would die without air, yet there is already in the box as great a quantity of air as could get in if there were dozens of holes. What you want is to give your pet fresh air from the outside so that it will not have to breathe over and over again that which becomes more poisoned every time the animal draws it into his lungs. When we shut ourselves up in rooms with tight win-

1. What is said of air that is not fresh? How is air altered every time it is breathed? What would happen if you tried to go on breathing the air shut up in a small bag? What is the real reason that an animal shut up in a box needs an air-hole? Apply to closed rooms.

dows and no open fireplaces, we are as badly off as the puppy would be in his box, without an air-hole, if you should occasionally open and shut the lid quickly, as we do our doors on a cold day.

2. Starvation and Suffocation Compared.—If a man gets no food, he soon dies of starvation. If he gets some food, but not enough for the needs of his body, he lives longer, but his whole body is weak. At last he dies of slow starvation, unless some disease attacks his feeble organs, and kills him before want of nourishment has had time to do so.

It is much the same as regards the supply of oxygen in the air we breathe. If there is no oxygen in it, death takes place in a few minutes. Death from suffocation occurs quicker than death from starvation, because our bodies have laid up in them but very little more oxygen than they need at the moment; whereas, in fat and some other tissues, there is a store of nourishing matter which the body can make use of when its food is not enough. Fat, when not present in such excess as to hamper various organs in their work, may be compared to a little money laid by in a savings-bank, and ready for use in case the regular supply gives out. There is no such bank in our bodies where extra oxygen can be stored. In health, the blood and each organ possess just a little more than they want at the moment, but that is all. It is like a few cents of pocket-money, which does not last long if we have to live on it.

2. What happens if a man does not get enough food? Of what does he die? If he gets no oxygen? Why does a man die of suffocation sooner than of starvation? To what may fat be compared? How much oxygen do the blood and organs possess in health? To what compared?

3. Foul Air is Worse than Insufficient Food.—If a man be nearly starved to death and then be carefully nourished, he may soon be all right again, but if he has been slowly poisoned, as well as starved, he is not so likely to recover. When a man does not get enough fresh air to breathe, he is not only starved for want of oxygen, but poisoned. The wastes or excretions of the foul air are, absorbed into the blood from the lungs, and are then carried by the blood to every organ. The health of the body, when pure air is breathed, depends on the perfection with which the blood carries oxygen to every nook and corner. When impure air is breathed, the hurtful substances taken into the blood as it flows through the lungs, are carried in exactly the same way to all parts.

You know that there are quick poisons and slow poisons. Quick poisons kill in a few minutes or a few hours. Slow poisons may not kill for weeks, months, or even years. Many slow poisons do not themselves actually cause death, but they so much weaken some of the organs that the body is very apt to take disease, and in its feeble condition cannot master and overcome it. Foul air is rarely foul enough to act as a quick poison, but unless proper care be taken, the air in rooms much lived in, soon becomes foul enough to act as a slow poison. How quick or how slow, depends simply on how foul the air may be.

3. What is the usual result if a man be carefully fed after being nearly starved to death? If he has been also poisoned? Apply to want of sufficient fresh air. How are the poisonous matters in foul air carried over the body?
What are quick poisons? Slow? How do many slow poisons act? What is said of foul air as a quick and a slow poison?

4. Rapid Death from Insufficient Supply of Fresh Air.—
Cases of quick poisoning from repeated breathing of the
same air are not frequent.   Fortunately, few doors and
windows fit so tight as to prevent fresh air from getting
into a room, and foul air out of it, fast enough to keep one
or two people alive.   The very deadly result of breathing
the same air repeatedly has, however, been terribly
proved in more than one instance.   The steamship
"Londonderry," a few years ago, sailed from Liverpool
with two hundred passengers on board.   Stormy weather
coming on, the captain ordered all the passengers into a
small cabin and then closed its openings.   " The wretched
passengers were now condemned to breathe over and
over again the same air.   This soon became intolerable.
There occurred a horrible scene of frenzy and violence,
amid the groans of the dying and the curses of the more
robust.   This was stopped by one of the men contriving
to force his way on deck, and to alarm the mate, who
was called to a fearful spectacle.   Seventy-two were
already dead and many were dying ; their bodies were
convulsed, the blood starting from their eyes, nostrils,
and ears."   All this occurred within six hours.

Not merely some fresh air, but a certain quantity of
fresh air is necessary to maintain life.   It seems almost
absurd to point out this fact, yet many folks act as if they
believed that any air-hole, with little regard to its size,
were sufficient.   The greater the number of people in a
room, the more abundant must the air-supply be.
Ignorance of this fact led to the horrible catastrophe of

4   Why is quick poisoning from foul air not frequent ?   Give an account of the example of it on board the " Londonderry."   What be
sides " some" fresh air is needful ?   When must the fresh air be more
abundant ?   Describe the catastrophe of the " black hole of Calcutta "

the " black hole of Calcutta." One hundred and forty-six prisoners were shut up in a small room with two narrow open windows. These windows would probably have supplied abundant fresh air for ten or twenty persons, but they were so insufficient for the needs of the large number locked up in the room, that, in eight hours, one hundred and twenty-three died.

5. **Ventilation.**—Most of us have to spend a large part of our time within more or less closed rooms. In order that the air in them may continue fit to breathe, it must be changed all the time. This removal of the foul air and its replacement by fresh, is known as *ventilation.* Ventilation is "sufficient" when it renews the air fast enough. It is *good* when, in addition to being sufficient, it does not cool a room too much or cause injurious draughts.

6. **The Amount of Ventilation Necessary** depends of course on many things. If there are two people living in a room, they will require just twice as much fresh air as one; and fifty will need fifty times as much. School-rooms, churches, theatres, and other like places, where many people collect, need very free ventilation. All such burning things as fires or candles or gas or oil-lamps, take valuable oxygen from the air and give hurtful carbonic acid to it. In ventilating a room, allowance must therefore be made for them. Ventilation just

5. What is necessary that the air in inhabited rooms may continue fit to breathe ? What is ventilation ? Sufficient ventilation ? Good ventilation ?

6. How does the number of persons in a room affect the amount of ventilation necessary ? Examples of rooms which especially need free ventilation ? How do burning things alter the air ? Why is more ventilation necessary when the gas is lighted ?

sufficient in the morning, will not be enough at night when the gas is lighted.

7. How Deficient Ventilation may be Recognized.—The nose generally affords the most sensitive as well as the most convenient test of the sufficiency of the ventilation of an inhabited room. If ill ventilated, the air will usually smell "close." Those who have been in the room for some time are not likely to realize how foul the air has become, as the nose gradually gets used to air around it, which would be extremely unpleasant to one just entering the room. If the room smells even the least bit "close" to a person entering it from out of doors, it needs more ventilation.

8. Consequences of Living in Insufficiently Ventilated Rooms.—A stay of an hour or two in a room not supplied with enough fresh air, results in headache, dulness, and sleepiness, which soon go off when we get out again into the fresh air. Children have often been punished for seeming neglect of their studies, when the foul air of the school-room was really to blame.

If one spends a considerable portion of every day in a badly ventilated room, the whole body is enfeebled. The blood becomes poor in red corpuscles, and the face pale; appetite is lessened, digestion imperfect, and the muscles weak. The body, not getting enough oxygen and being at the same time slowly poisoned by breathing its own wastes over and over, has but little reserve force. It is

7. How does the air of an ill-ventilated room affect the nose? Why may foul air not be perceived by those who have been some time in an insufficiently ventilated room? When does a room need more ventilation?

8. What are the consequences of staying for an hour or two in a badly ventilated room? What of spending several hours daily in

liable to take disease, and when disease occurs, there is less chance of recovery.

Consumption and other lung-diseases are especially frequent in persons who live in badly ventilated rooms. So are colds of all kinds.

9. Free Chest-Movements are Necessary for Healthy Breathing.—Plenty of fresh air to breathe is not of much use if the chest is so imprisoned that it cannot expand properly. No garment which checks the free movements of thorax and abdomen in breathing, should be worn. The tight lacing which used to be thought elegant, and is still indulged in by some who think a distorted form beautiful, does harm in many ways. In the first place it makes all healthy exercise impossible. A tightly laced person gets "out of breath" on the least exertion. Many a woman complains that she is unable to attend to her household duties, because the least exertion fatigues her, when all that is the matter is that she has so laced her chest that it cannot do its breathing work properly. Tight lacing also hampers the abdominal organs. It so narrows the chest (Fig. 12) that lungs and heart are pushed down towards the abdomen, to get room. The heart is driven so close against the stomach that even a moderate meal is apt to press unnaturally against it (p. 127), and so its working is interfered with. The livers of those who have practised tight lacing are often found to have hard unhealthy cords on them, caused by pressure from the lower ribs, squeezed in by the corset.

badly ventilated rooms? What diseases are especially frequent in those who live in ill-ventilated rooms?

9. What is necessary for healthy breathing, besides pure air? How does tight lacing do harm as regards exercise and work? As regards the heart? As regards the liver?

10. Expansion of the Chest by Exercise.—Some persons
are born with narrow chests, and are predisposed to lung-
diseases. Proper exercise, regularly performed, will do
a great deal to widen the chest. Rowing is good for

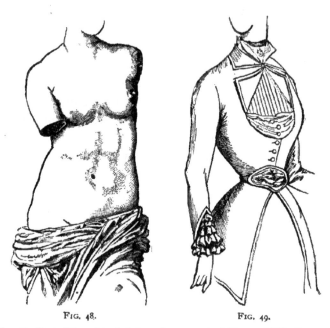

FIG. 48.   FIG. 49.

FIG. 48.—Part of the celebrated statue known as the " Venus of Milo," a recognized standard of female beauty.
FIG. 49.—The dressmaker's idea of a beautiful waist.

this purpose, but certain gymnastic exercises are better.
They often increase the size of the thorax even in a few
weeks. A delicate person should get skilled advice as to
the kind and amount of work to do in a gymnasium.
Otherwise he may easily do himself harm.

10. How may the chest be made larger? What should a delicate
person do before beginning work in a gymnasium? Why?

11. Mouth-Breathing.—Quite a number of people breathe through the mouth instead of the nose. This not only gives the face a weak silly look, but it tends to cause disease of the lungs and air-passages.

When air is breathed through the nose, it has to pass through a long narrow passage lined with warm moist mucous membrane, before it gets into the pharynx. In this way it is warmed and moistened before it enters the larynx, on its way to the lungs. Air breathed in through the mouth is apt to be too cold or too dry when it reaches the bronchial tubes, and to injure them and the air-cells of the lung.

The nostrils are very often blocked during a cold in the head, but if your nostrils are usually so stopped that you find difficulty in breathing through them they should be examined by a physician, in order that whatever causes the stoppage may be removed. If a child habitually breathes through the mouth when asleep, it is probable that something is wrong with its nose.

12. Action of Alcoholic Drinks on the Respiratory Organs.—Indulgence in alcoholic drinks often keeps the mucous membrane lining the air-passages in a congested state. It thus increases the tendency to colds of the head and chest. There is also a peculiar form of consumption of the lungs, which is rapidly fatal, and is found only in drunkards.

11. What must air, when breathed through the nose, do before it reaches the pharynx? What results? Why is air breathed in through the mouth likely to injure the lungs? What should be done if you have continual difficulty in breathing through the nose?

12. Action of alcohol on the air-passages? Results? What lung-disease is found specially in drunkards?

CHAPTER XVII.

THE KIDNEYS AND THEIR FUNCTION.

1. Why the Kidneys are Needed.—We have seen how the body gets rid of one of its chief waste matters, namely, carbonic acid. Another waste substance is formed in it every day in large quantity, and if not carried out would do just as much harm as carbonic acid. This waste substance is named *urea*. It is solid, and so cannot be separated by the lungs, which can pass out gases and vapors. The urea is removed by the *kidneys*, along with a great deal of water in which it is dissolved; it is thus passed out in a liquid form.

Urea contains nitrogen, and is produced when albumens are oxidized (p. 83), or used up, in doing their work in the body.

2. The Renal Organs include not merely the kidneys, but the apparatus by which their secretion is carried to the outside of the body and expelled from it. They are: (1) the kidneys, two large glands placed in the abdomen; (2) the *ureters*, or the ducts (p. 66), of the kidneys, which carry the secretion to (3) a reservoir, the *bladder*, where it collects. The bladder is a muscular bag. It contracts

1. What is urea? Why can it not be separated by the lungs? What organs remove it? In what form? What does urea contain? How is it produced?
2. What do the renal organs include? Name them. Function of ureters? Of bladder? Of urethra? When do the kidneys work?

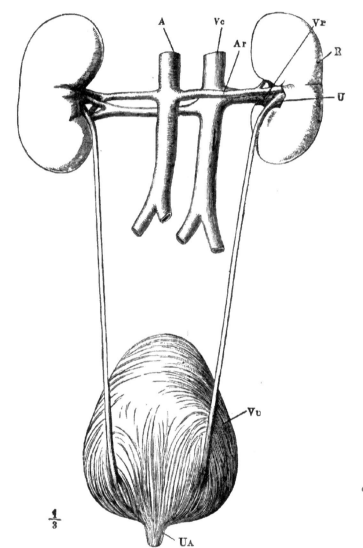

FIG. 50.—The renal organs, one-third life size, viewed from behind. *A*, lower end of aorta; *Ar*, the right renal artery; *R*, the right kidney; *U*, the right ureter; *Vu*, the bladder; *Ua*, commencement of urethra; *Vr*, lower end of inferior vena cava; *Vr*, the right renal vein.

from time to time and expels the liquid which has gathered in it, through a passage (4) named the *urethra.* The kidneys are at work all the time, separating urea from the blood, though the bladder only empties out their secretion a few times a day.

 3. The Kidneys lie at the back of the abdominal cavity, on the sides of the vertebral column, a short way below the diaphragm. Each is about half as big as its owner's clenched fist. The blood is sent to the kidneys for purification by two large branches of the aorta, named the *renal arteries.* The kidneys not only take urea from the blood, but help in removing other waste matters.

 4. The Chief Excretory Organs Compared as to their Functions.—*The skin* gets rid of a good deal of water, of some mineral matters which have done their work, and sometimes of a little urea. The duties of the skin as an excretory organ are important, and health cannot be maintained if they are badly performed. But the chief functions of the skin are to protect deeper parts, to regulate the temperature of the body (p. 77), and to give us the sense or feeling of *touch* (Chap. XXI.).

 The lungs get rid of much carbonic acid, of small quantities of very poisonous animal vapors, and of some water. They separate no mineral wastes and no urea. The function of the lungs as receptive organs, to supply the body with oxygen, is as important as their excretory function.

 The kidneys are solely excretory organs. To get rid of

3. Position of the kidneys? Size? How is blood carried to them? What do they do besides taking urea from the blood?
4. What is said of the skin as an excretory organ? Of its other functions? Of the lungs as excretory organs? Of their other duty? Of the duty of the kidneys? What do the kidneys remove from the body?

waste matters which would poison the body is their only duty. Except a very little carried off by the skin, they remove all the waste matters containing nitrogen, a great deal of water, nearly all the mineral wastes, and some carbonic acid.

5. Hygiene of the Kidneys —If both kidneys be cut out of an animal, it dies in a few hours from blood-poisoning, caused by the wastes which have collected in it. Serious kidney-disease amounts to pretty much the same thing as cutting out the organs, since they are of little use if not healthy. It is always fatal if not checked, and often kills in a short time. The things which most frequently cause kidney-disease are undue exposure to cold, and indulgence in alcoholic drinks.

6. Cold Causes Kidney-Disease partly by driving blood from the surface and congesting the kidneys, partly by throwing too much work on them. When the skin does not get rid of its proper share of the waste matters of the body, it is chiefly the kidneys which have to make up for it.

Nearly all the infectious diseases which are accompanied by a rash on the skin, as measles and scarlet fever, also affect the kidneys. During these diseases, the kidneys are more or less inflamed, and in the early stages of recovery they are still weak and easily injured. Under these circumstances, exposure to cold is very apt to cause incurable kidney-disease.

5. What is the consequence of removing the kidneys? Of kidney-disease? How is serious disease of the kidneys most often produced?
6. How does cold injure the kidneys? When have they to do the work of the skin? What diseases especially affect them? State of the kidneys during recovery from these diseases? Precautions to be taken?

7. Alcohol Causes Kidney-Disease in Several Ways.—In the first place it overstimulates the organs. Next, when its abuse is continued, it interferes with the proper preparation of the nitrogen wastes: they are then brought to the kidneys in an unfit state for removal, and injure those organs. Third, when more than a small quantity of alcohol is taken, some of it is passed out of the body unchanged, through the kidneys, and injures their substance.

The kidney-disease most commonly produced by alcohol, is one kind of "Bright's disease," so called from the physician who first described it. The connective tissue of the organ grows in excess, and the true excreting kidney-substance dwindles away. At last the organ becomes quite unable to do its work, and death results.

7. State one effect of alcohol on the kidneys. Another? A third? What kidney-disease is commonly produced by alcoholic excess? How are the kidneys altered by it? Results?

APPENDIX TO CHAPTER XVII.

To demonstrate the anatomy of the renal organs proceed as follows:

1. Kill a rat, puppy, or kitten in any merciful way; placing it under a bell-jar with a sponge soaked in ether is a good method.

2. Open the abdomen of the animal, remove its alimentary canal, and cut away (with stout scissors) the front of the pelvic girdle. The dark red *kidneys* will then be easily recognized on each side of the dorsal part of the abdominal cavity, the right one nearer the head than the left.

3. Dissect away neatly the connective tissue, etc., in front of the vertebral column, so as to clean the *inferior vena cava* and the *abdominal aorta*. Trace out the *renal arteries* and *veins*.

4. Find the *ureter*, a slender tube passing back from the kidney towards the pelvis. it leaves the inner border of the kidney behind the vein and artery; and lying, at first, at some distance from the middle line, converges towards its fellow as it passes back

5. Follow the ureters back until they reach the *urinary bladder;* dissect away the tissues around the latter and note its form, etc.

6. Open the bladder; find the apertures of entry of the ureters, and pass bristles through them into those tubes. Note the *mucous membrane* lining the bladder.

CHAPTER XVIII.

THE NERVOUS SYSTEM AND ITS FUNCTIONS.

1. Introductory.—If the inside of your nose be tickled, you cannot help sneezing; it seems so natural to sneeze when anything irritates the nostrils that probably you never thought about it at all. But if you do think about it, you will find that it is something quite curious and interesting. If some one puts a soft feather up your nose, neither the larynx, nor the lungs, nor the chest-muscles, nor the diaphragm, are interfered with; yet they all (p. 174) set to work at once to help the nose to get rid of what is worrying it, and they do this without paying any heed to your will. In other words, they act *involuntarily.* They do, apparently of themselves, what is likely to help the nose, and they set to work in a very orderly way. If any one of them failed to do its share of the work, or worked never so little out of its turn, no useful sneeze would be produced.

How the nose obtains such ready and well-planned help from all these organs which lie at a distance from it, we will try in this chapter to explain.

2. Other Examples of the Help which our Organs give to One Another.—Coughing (p. 174) is one that will of

---

1. What results from irritating the inside of the nostrils? What organs work together to produce a sneeze? What is meant by saying that they act "involuntarily"? What would happen if any one of them did not act "just right"?

course come to your mind at once. There are others
that you may think of, as you have also learned that
when you exercise your muscles, the heart and lungs
work more vigorously to supply them with sufficient
nourishment and oxygen, and to carry off their extra
wastes; that when the air is cold, the blood-vessels of
the skin contract and drive blood away from the surface
to prevent too rapid cooling (p. 178); that when your
body is hot, the sweat-glands become very active so as to
cool the blood, and through it the internal organs (p. 67);
that when partly digested food passes from the stomach
into the small intestine, the gall-bladder at once squeezes
out bile (p. 116) to be mixed with it, and help the intestine in digestion and absorption.

All of the things above mentioned are done without
our will, and some of them even without our being
aware when they take place, as the pouring of bile into
the small intestine. They are but a few examples out
of hundreds, which show that our organs work together
for the good of the whole body, and often help one another without our planning it, or our minds having anything to do with it. Very clearly there must be some
means by which the various organs are made to work in
such harmony.

3. **The Nervous System.**—When we try to imagine how
each organ might be put in communication with all the
others, probably the first idea that comes to mind is that

2. Of what is coughing an example? How do the heart and lungs
help the muscles during exercise? How do the blood-vessels of the
skin keep the rest of the body from being too much cooled? How do
the sweat-glands aid the rest of the body? How does the gall-bladder
aid the small intestine in digesting? In what way are the above actions performed? What do these few examples show?

3 In thinking of communication between the organs, what idea

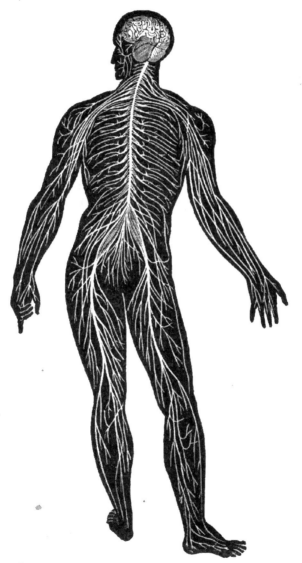

Fɪɢ. 51.—Diagram illustrating the general arrangement of the nervous system.

there might be some sort of telegraph-system in the body. If there were something like telegraph-wires running from all the organs to a central office or exchange, then word of the state and needs of any organ might be sent from it to the central office, and proper messages be sent out from the central office to those other organs whose help was wanted. This is in fact something very like what does take place.

If the dead body be dissected, a great many white cords are found which run all through it, and go into the skin, and the mucous membranes, and the heart, and the lungs, and each muscle, and so forth. These cords are *nerves.* If one be followed back from where it enters any of the above parts, it will be found at last to join a much larger mass to which other nerves are also united. This mass is a *nerve-centre.* The nerves and nerve-centres together make the *nervous system.* The nerves answer to the telegraph-wires, and the centres to the main offices from which the wires spread over the country.

4. The Chief Nerve-Centres are the brain, the spinal cord, and the *sympathetic ganglia.* You have already learned that the brain lies inside the skull (p. 19), and the spinal cord runs down inside the back-bone. At the under part of the skull, where it fits on the back-bone, is a large hole, through which the brain and spinal cord unite. Strictly speaking, therefore, the brain and spinal cord make only one centre; they are often spoken of

might occur to us? What really does take place? What are nerves? What is found when a nerve is traced back from a muscle or the skin? Name of the mass? Of what does the nervous system consist? To what are nerves and nerve-centres compared?

4 What are the chief nerve-centres? Where does each lie? How do they join? What is the cerebro-spinal centre?

together as the *cerebro-spinal centre*. The sympathetic ganglia will be described farther on.

5. The Brain of an adult usually weighs about three pounds.

It has two chief parts (Fig. 52), the *great brain* or *cerebrum, A,* and the *small brain* or *cerebellum, B.* It is joined to the spinal cord by the *medulla oblongata, D.* The parts

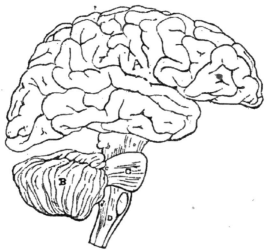

FIG. 52.—Diagram illustrating the general relationships of the parts of the brain as seen from the side. *A*, cerebrum ; *B*, cerebellum ; *D*, medulla oblongata.

of the brain are not really so widely separated as is represented, for the sake of clearness, in Fig. 52. They lie closely packed together, as shown in Fig. 53.

The cerebrum fills all the front and upper part of the skull-cavity. It is much larger than the cerebellum, and

5. What does the brain of a grown person usually weigh ? What are its chief parts ? How joined to the spinal cord ? How do they lie in the skull ? Relative size ? What are the cerebral hemispheres ? How are their surfaces marked ? Name of the ridges ?

its hinder end laps over it. A deep groove runs along the cerebrum from front to back and nearly cuts it in two. Its halves are named the right and left *cerebral hemispheres*, and their surfaces are not smooth but are marked by numerous crooked furrows, with ridges between (Fig. 53). The ridges are known as the *convolutions*.

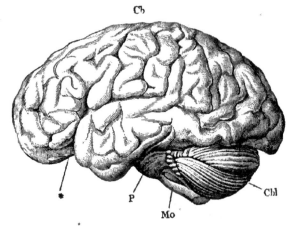

FIG. 53.—The brain from the left side. *Cb*, the cerebrum, or, rather, the left cerebral hemisphere; *Cbl*, the cerebellum; *Mo*, the medulla oblongata.

6. The Spinal Cord is nearly round, and is about three quarters of an inch across and seventeen inches long. It does not reach as far as the lower end of the back-bone.

7. The Nerves start from the brain and spinal cord. Twelve pairs (*cranial nerves*) are attached to the brain and go out through holes in the skull; thirty-one pairs (*spinal nerves*) spring from the sides of the spinal cord, and pass out between the vertebræ.

6. Describe the spinal cord. How far does it reach?
7. Whence do the nerves start? What is said of the cranial nerves? Of the spinal? Of the nerve-fibres? Describe the branching of nerves,

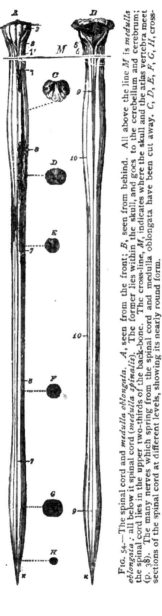

FIG. 54.—The spinal cord and *medulla oblongata. A,* seen from the front; *B,* seen from behind. All above the line *M* is *medulla oblongata ; all below it spinal cord (medulla spinalis).* The former lies within the skull, and goes to the cerebellum and cerebrum; the spinal cord lies in the upper two-thirds of the back-bone. The cross-line, *M,* indicates where the skull and the atlas vertebra meet (p. 38). The many nerves which spring from the spinal cord and medulla oblongata have been cut away. *C, D, E, F, G, H,* cross-sections of the spinal cord at different levels, showing its nearly round form.

Each nerve is made up of a number of very slender threads, named *nerve-fibres,* which run side by side in it like the threads in a skein of silk. As a nerve is followed along from its centre, it is found that it separates into smaller bundles of fibres, which run off as branches. These branches again divide, and so on, until the last branches are very small and very numerous.

8. Sensory and Motor Nerve-Fibres.—A telegraph-wire is used to send messages both ways. The same wire will carry a message just as readily from New York to Chicago as from Chicago to New York. Our nerve-fibres are not used in this way. Some of them are always employed to carry messages *to* the centres, others to carry messages *from* the centres. The fibres

8. Point out an important difference between the carrying of messages by telegraph-wires and by nerve-fibres. What is meant by sensory fibres? By motor?

which carry towards a centre are usually called *sensory* fibres, because when they work they very often cause some sensation or feeling. The fibres which carry from a centre are named *motor* fibres, because they usually cause some muscle to contract, and thus produce movement. The first set of nerves is also sometimes called *afferent* (from a Latin word meaning bringing to), and the other set *efferent* (from a Latin word meaning to bring from). These names are better than sensory and motor, because many nerves carry messages to centres without our having any sensation of them, and many nerves carry messages from centres to other organs than muscles, for example to glands.

9. Reflex Movements.—As we have seen, a great deal of the orderly working of our organs is brought about without our will, or even without our knowing about it. When a message comes to a nerve-centre, the centre does not merely send out random messages along any outgoing nerve-fibres, but, as it were, first selects the organs to be set at work, and then sends the proper messages. As, for instance, in the case of sneezing. If the centre, warned by the sensory nerves of the nose, should set at work any or every outgoing fibre joined to it, the result would not be a sneeze, but some sort of a shaking-up or *convulsion* of the whole body. It might once in a thousand times be useful, but in most cases would do more harm than good. The disease known as " convulsions" is due to the fact that the nerve-centres, whenever a nerve-fibre brings a message to them, send out random

9. How is the involuntary working of our organs managed by the nerve-centres? Illustrate from the case of sneezing. What happens if the nerve centres send out messages to the wrong organs? To

messages to all the muscles instead of only to those whose contraction would be useful.

Nerves merely carry messages to and fro. Nerve-centres do much more than this; they guide the messages to the various organs, and, in all ordinary circumstances, make them work for the general welfare of the body. Most nerve-centres do this independently of our will; they set the proper muscles at work whether we like it or not, though the cerebrum, which is the largest nerve-centre and only one where the will acts, sends out most of its messages in answer to the will. When food goes the wrong way and gets into your larynx (p. 111) you cannot help coughing; when something comes rapidly close up to your eye you cannot help winking; when you chew food you cannot prevent your salivary glands (p. 109) from pouring out extra secretion. All such useful movements, guided by nerve-centres, and not dependent on our will, are known as *reflex movements.* Sometimes we notice them, though we cannot hinder them, but far oftener we know nothing about them. These unconscious reflex movements, guided by the nerve-centres, carry on nearly all the regular daily work of the body necessary to keep it alive. They regulate the circulation and the breathing, and the secretion of the digestive liquids, and so on. The medulla oblongata especially regulates the beat of the heart and the breathing movements; if it is seriously injured, death occurs very quickly.

If we had to think about and will every beat of the

what are "convulsions" due? What is the function of nerves? Of nerve-centres? How do most nerve-centres behave as regards our will? Illustrate. Explain what is meant by reflex movements. What is said of our consciousness of them? Of their use?

heart, and the drawing of every breath, and the secretion of digestive fluids in the proper amount at the proper moment, and the blood-flow through each organ according to its needs at that time, and so forth, our minds would have time for nothing else. All this daily routine is looked after by nerve-centres which act involuntarily, and leave the mind free for other duties.

10. Feeling and Willing.—The spinal cord, the medulla oblongata, and the cerebellum direct unconscious and involuntary movements. The cerebrum guides some such movements, but it does more: it is connected in some way with feeling and willing. No part of the body which is not joined by at least one nerve-fibre to the cerebrum, has feeling ; and no muscle not joined to it in like way, can be controlled by the will.

For example, the nerve-fibres coming from the leg all unite, above the hip, into three or four large cords, which enter the spinal cord near its lower end. If all the nerves be cut at the ankle, the foot loses feeling, and all the muscles in it are *paralyzed;* that is to say, cannot be made to contract by their owner when he wishes. If only some of the nerve-fibres going to the foot be cut, then only that part of it to which the divided fibres went, loses feeling and has its muscles paralyzed. If all the nerves be cut at the knee, instead of the ankle, then both the foot and the lower part of the leg become insensible and paralyzed. If they be divided or crushed at the hip-, joint, then the thigh also is put in the same condition.

10. What centres direct most involuntary movements ? What part of the body is especially concerned in feeling and willing ? What is said of muscles and other parts not joined to the cerebrum by a nerve-fibre ? Illustrate from the results of injuries to the nerves of the leg at the ankle. The knee. The hip. What results when the spinal

If the nerves of the leg be not injured at all, but the spinal cord be cut or seriously diseased above the place where they join it, the leg loses all feeling and has its muscles paralyzed just as if its nerves themselves were cut. The reason of this is tha* ..ie nerve-fibres which run up the spinal cord to the cerebrum and cause feeling, and those which run down from the cerebrum to the leg and make its muscles obey the will, have been divided. The spinal cord, in addition to being a centre itself for many reflex movements, is a sort of nerve : it affords a path for many nerve-fibres which run between the cerebrum and most parts of the body.

11. The Sympathetic Nervous System.—In addition to the great system of nerves we have been studying, which branches out from the brain and spinal cord, and then divides and divides until it reaches every organ, and covers the surface of the body as closely as the capillaries (p. 147) do, so that the prick of a pin-point must touch one of the little branches—in addition to this great set of *cerebro-spinal nerves* there is another, called the *sympathetic system.* The nerves of the sympathetic system are not spread through the skin or concerned in the sense of touch; nor are they subject to the will and concerned in producing voluntary movements. But they go to the lungs, and the heart, and the liver, and the stomach and intestines, and to the involuntary muscles (p. 42). They do not run direct to the brain and spinal cord, but first to certain smaller centres, lying principally

cord is cut above where the nerves of the leg enter it ? Why ? · What is the spinal cord in addition to being a centre ?
11. What is said of the nerves connected with brain and spinal cord ? Of the nerves of the sympathetic system ? What is a ganglion ? Why so named ? What is the sympathetic system ? Its duties ?

in two rows in front of the spinal column (*s*, Fig. 1).
Each of these small centres is named a *ganglion*, which is
the Greek word for a swelling. This name has been
given them because they make swellings on nerves like
knots on a string. These *ganglia* are joined to one another and to the brain and spinal cord by nerves. They,
with the nerves running to and from them, look after a
good many of the details of the working of the body.
The sympathetic system is a sort of under-servant of the
brain and spinal cord, trusted to look after certain routine
work, especially the distribution of the blood among the
various organs, according as their needs may be. It has
also much to do with managing the glands. It owes
its name to the fact that it makes many organs which
are not under direct control from the will, work together
as if they sympathized with one another.

12. Mind and Brain.—The cerebrum is not only concerned in feeling and willing, but in remembering and
reasoning, and in all the other things which go to make
up what we call mind and character. How mind is connected with brain it is not possible to imagine ; we have
just to accept the fact that it is, and especially with its fur-
rowed and ridged surface. When this is seriously diseased, feeling is lost or unnatural, the will is enfeebled,
memory weakened, reason impaired, and the man no
longer capable of judging correctly, nor really responsible
for his actions. Why, or how, we do not know, and

12. With what besides feeling and willing is the cerebrum concerned ? What is said of the connection of cerebrum and mind ?
What part of the cerebrum has especially to do with mind ? What is
seen when it is seriously diseased ? What is it sufficient to know
concerning the connection of brain and mind, for all practical purposes ?

probably never will know. However, for all practical purposes, it is sufficient to know that, if we desire active and vigorous minds, we must try to keep healthy brains; we may then consider all the knowledge we can get about the hygiene of the brain as coming, in the long-run, to the same thing as hygiene of the mind.

CHAPTER XIX.

HYGIENE OF THE NERVOUS SYSTEM.

1. Introductory.—The nervous system is so closely connected with all other parts of the body that anything which injures them can hardly fail to hurt it. He who desires an active healthy nervous system and a vigorous cheerful mind, must strive to keep muscles and digestive, circulatory, and respiratory organs in health.

On the other hand it should be borne in mind, that nearly every function of the body is dependent on the nervous system for its proper performance. It sets at work the muscles which move the jaws, and the glands which secrete saliva; controls the œsophagus in swallowing; excites the glands of the stomach, and makes its muscular coat mix the gastric juice with the food; it governs the secretion of pancreatic juice and bile, which turn the chyme into chyle; makes the muscular coat of the intestine drive the digesting mass along that tube, and controls absorption by its lacteals and blood-vessels; it regulates the beat of the heart, and the diameter of the arteries, and, thus, the blood-flow to every organ; it

---

1. What is said of the connection of the nervous system with other parts of the body? What must one do who desires an active nervous system and mind? What should also be borne in mind? Give illustrations of the action of the nervous system in preparing food to enter the stomach. In controlling its digestion in the stomach. On conversion of chyme into chyle? On the movements of the intestine? On absorption? On the blood-flow? On excretion?

keeps in action the lungs, and skin, and kidneys to purify the blood; it makes the eye see and the ear hear; and through it we think, and hope, and love. To injure the nervous system by too much work, too little sleep, or over-indulgence in tobacco, alcohol, or any other substance which affects it, is to weaken every function of the body and the mind.

No doubt many persons have attained intellectual eminence and led happy and useful lives in spite of bodily feebleness. Unusual strength of will has enabled them to overcome the odds against them. But we should remember that body and mind are so united that any disease of one affects the other, and should guide our conduct accordingly.

2. Some Disorders of the Nervous System.—Unhappily most children have seen cases of " *St. Vitus' Dance.*" It is a twitching of the muscles, sometimes only those of the face, sometimes those of the limbs and body generally. It comes from weakening of the control of the nervous system over the muscles, so that occasionally some muscle relaxes. This enables the opposing muscle to give a jerk and pull the organ, it may be the eyelid, the mouth, the arm, or the leg, out of place. Sometimes these jerkings are so violent as to seriously injure the organs.

Fit is a name given to several disorders attended with loss of consciousness. A *fainting fit* is due to temporary weakness of the heart; it pumps so little blood around that the cerebrum does not get enough nourishment to

What are the consequences of injuring the nervous system? What have some persons of feeble body accomplished? How? What should we remember?

2. What is St. Vitus' dance? To what due? To what is a fainting

enable it to work. A person who has fainted should be
laid at once flat on the back, with the head low; this
enables blood to be pumped more easily to the brain.
The skin may then be stimulated by sprinkling the face
briskly with cold water, or the nose by holding harts-
horn to the nostrils. The *convulsions* so common among
infants are in most cases excited by some irritation connected with the alimentary canal. An emetic should be
given at once, cold applied to the head, and the body
put in a warm bath. In *epileptic fits* there is usually a
peculiar cry, the face becomes pale, consciousness is
lost, and then convulsions (p. 197) occur. Lay the person flat, and restrain any of his movements likely to injure him. If possible, a folded handkerchief should be
pushed between the teeth to prevent biting of the
tongue. After the convulsions have ceased, quiet is desirable. *Hysterical fits* assume many different forms, the
more frequent perhaps being unreasonable screaming,
laughing, and weeping by turns. They should be
noticed as little as possible. A display of interest and
sympathy nearly always makes a fit of hysterics last
longer. A little rudeness, exciting anger, is often the
best treatment.

An *apoplectic fit* or *apoplexy* is due to the bursting of
some blood-vessel of the brain. The blood which flows
out compresses the brain, and the person becomes more
or less unconscious. The breathing is heavy and like
snoring, and the face usually flushed. A person suffering from an apoplectic fit should not be moved at all if

fit due? Treatment. What is said of the convulsions of young children? Characters of an epileptic fit? What is said of hysterical fits?
Cause of an apoplectic fit? Symptoms? What is said of the manage-
ment of a person in an apoplectic fit? What is neuralgia? On what

it can be avoided; apply cold to the head until medical aid can be obtained.

Neuralgia is a diseased condition attended with intense pain, which may attack almost any part of the body. It seems to depend on an altered or disordered state of the nerves themselves, for usually nothing can be found wrong in the organ in which the pain is felt. Thus the teeth or the stomach may appear to be perfectly sound in their structure, and yet suffer intensely from neuralgia. The almost unbearable pain often leads to the use of alcohol, opium, and chloral (Chap. XX.), drugs which, while giving temporary relief, tend to increase the diseased condition of the nerves. Some persons have organizations more nervous than those of others, and under unfavorable conditions of life are very apt to become victims of neuralgia. These persons may be recognized by their tendency to undertake more than they have the strength to perform safely, and to be extreme in all their feelings. They should guard against lives of excitement, and be careful to secure plenty of sleep, and not to allow themselves to be overdriven by ambition.

3. The Three Great Sources of Nervous Health are a brave heart, a cheerful disposition, and plenty of sleep.

Nothing wears the nerves like worry. The child at school who keeps a brave heart for whatever may happen stands a better chance of success than the one who wears his nerves out with constant dread of failure. One

does it depend? Illustrate. To what does it often lead? How may persons apt to become neuralgic be recognized? What precautions should they take?

3 What are the three great sources of nervous health? What is said of worry? Of the effect of a brave heart in promoting success? Of the benefits of a cheerful disposition?

who has a cheerful disposition and a sunny temper is not only unlikely himself to suffer from nervous ailments, but, by a contagious influence, helps to keep others well and happy.

4. Sleep, however, sound and plenty of it, is the one great condition of nervous health. The use of sleep is to give the cerebrum a period of complete rest, for growth and repair. While awake, even when we are not doing brain-work, the mind and cerebrum are in action all the time; feeling and willing and thinking. Perhaps not feeling much or willing much or thinking hard, but still doing some or all of those things every moment. So long as we are conscious, the mind and cerebrum are at work. Healthy sound sleep is a state of the body in which the cerebrum is entirely at rest and there is no conscious-ness. A due amount of it is as absolutely necessary for a healthy brain and mind, as periods of rest are for the muscles or stomach.

5 The Amount of Sleep Necessary for Health varies with age and employment. Children need more sleep than older persons, and those whose chief work is men-tal, need more than those whose work is muscular.

The brain of a child has to grow and develop and is easily fatigued; it needs plenty of the deep thorough rest given by sleep. Moreover the muscles of a healthy boy or girl are full of life, and need abundant exercise. This makes severe mental work dangerous (p. 57). The organs which nourish the body, can only in a few favored persons provide at the same time for the needs

4 What is the use of sleep? What is said concerning mind and cerebrum during waking hours? What is healthy sound sleep?
5. What persons need most sleep? First reason why children need more than adults? Another reason? What is the usual result of

of active growing muscles and hard-worked nervous systems. The attempt to make them do so, is very apt to stunt and injure both. As we grow older, and the demands of the body for extra materials for its growth become less or cease, more steady and continued brain-work can be undertaken with safety and benefit.

The "soundness" of the sleep is important. Five or six hours of thorough deep sleep, with no dreams or consciousness of any kind, are better than eight or nine hours of uneasy sleep. Sleepnessness (*insomnia*) is a very serious matter; if continued or frequent, medical advice should be obtained. Unless checked, it leads to exhaustion of the brain, and impairment of the mind.

6. The Brain Needs Exercise.—If the body in general is healthy, the involuntary nerve-centres will look after their own work, and take proper exercise and rest ; but the part of the brain concerned with mental work is more under our control, and may be harmed by over-work or idleness. It is made stronger, and the mind more vigorous, by regular exercise.

When one first begins to train his muscles to do any special task, they soon tire, but after a time the work becomes easy, and more difficult feats can be under-taken. In like way, mental work is apt at first to be very fatiguing, but regularly repeated, with proper intervals of rest, it becomes easier every time; and soon harder tasks can be accomplished, and even enjoyed.

trying to work hard with both brain and muscles? What is said of sound sleep as compared with restless ? Of sleeplessness?

6. What do the involuntary nerve-centres do in health? What part of the nervous system is more in our control? What is the effect of exercise on the mind? What is the result of training the muscles? The mind? What is said of the effects of idleness on the mind?

An idle mind, like idle muscles, becomes weak. Even if it remain in a few cases shrewd and clear, it is incapable of prolonged steady effort, such as may any day become necessary. There are mental loungers as well as muscular; and the former are rather the more contemptible.

7. Mental Exercise should be Varied.—You have learned (p. 57) that a man may exercise and greatly develop some of his muscles, and leave others idle and feeble. A great many people do something of this kind with their brains. They use and train some mental faculties and leave the rest unemployed until they almost cease to be active at all. The hard struggle which most of us have, nowadays, to make a place for ourselves in the world and keep it, is very apt to lead to this mental lopsidedness, which is as much a deformity as would be huge arms and spindling legs on the same body. We meet business-men so absorbed in money-getting that they care for no books except ledgers, no science unless it helps them to patent some invention. We meet men of science who take no interest in art or literature, or who affect to despise the business-men who are carrying on the great commerce which promotes the progress of the world in ten thousand ways. We meet literary men who seem quite incapable of sympathy with science, and artists who care for nothing outside of art. All such people may be very far from insane, in the usual sense of the word, but they are all mentally deformed.

7. How do some people train their mental faculties? What often leads to mental lopsidedness? To what is it compared? Give illustrations of persons who use only a small part of their mental faculties. What is said of mental deformity? Why is a broad education in early life very valuable?

Some are born so and cannot help it, but a great many have made themselves so by persistently neglecting to use many of their intellectual faculties.

After a man gets settled down to his business, whatever it be, he rarely has much time or energy to devote to other things.   Hence arises the value of a broad education in early life, tending to widen the range of our sympathies and interests.

8. Education.—All education worthy the name, not merely supplies instruction in certain things useful to know, but trains the will and strengthens the character. For this reason it should include the performance of unpleasant or difficult duties.   Every man and woman has to face many such duties in the course of life, and the will must be made strong to meet them.   A school where every study is made easy and pleasant may be popular, but it is not the best school to turn out real men and women, strong to play a noble part in life.

9. The Brain Needs Rest as well as Work—Overwork, giving no sufficient periods of rest for repair of the nerve-substance destroyed during activity, harms the brain very much in the same way as it does the muscles (p. 52).   The results of mental overwork are, however, apt to be far more disastrous than those of muscular. Muscles which have been exercised too much usually recover completely with rest and nourishment, and become as strong as ever: a brain which has given way under overwork, is very apt never again to be as capable of

8 What does all good education do?   Why should it include difficult tasks?   Why must the will be made strong?

9. What is said of overwork of the brain?   Why worse than of the muscles?   What are the mental symptoms of an overtaxed brain? How is the body in general affected by it?   How does it often lead to drunkenness?

continued labor as it would have remained, had it been used wisely.

Apart from mental symptoms, as sleeplessness, confusion of thought, low spirits, loss of memory, and incapacity for prolonged steady thought, an overtaxed brain acts on the whole body and injures it. The digestion especially is impaired, and this of course brings in its train many evils, due to ill-nourishment of various organs (see pp. 123–4). The feeling of lassitude and exhaustion causes a longing for stimulants, which give temporary relief, and many a man has thus become a drunkard.

10. Brain-Rest Obtained by Change of Employment — There is an old saying that "change of employment is as good as rest;" properly understood it is a very true one. The change, however, must be thorough. It is not of much use for a business-man to go, in search of rest, from New York to Saratoga and there continue his business by correspondence; nor for a child to change from studying history to arithmetic. Unless the change is accompanied by a sense of recreation and pleasure, it is of little or no value as affording brain-rest. Doing nothing is often wearisome to persons who have never formed habits of idleness; when the minds of such need rest, they should seek some occupation calling for little exercise of the faculties employed in their regular daily work, and which yet interests and amuses them.

11. Concentrating One's Thoughts.— One of the hardest things a child has to learn, is to "fix its attention," or

10. What is necessary that change of employment may rest the mind? Illustrate. What should accompany the change? What should those seek who soon weary of doing nothing and yet need brain-rest?

11. What is said of fixing the attention? Illustrate How may the power be acquired? Why should the training be gradual?

keep its mind from being distracted and wandering off to other things. A great many grown people, indeed, cannot do it. A very distinguished American lecturer, writer, and anatomist,* has stated that he could gauge the intelligence of his audience by the way in which they behaved when any slight disturbance occurred during his lecture. On an educated audience, with trained power of attention, any slight noise had little influence, while less educated hearers turned their heads at every trivial interruption.

To acquire this power of attention, is most important. Probably no young healthy child has it; it must be gained by prolonged training, but the training should be gradual. A young child cannot fix its mind on a lesson, no matter how easy, for an hour at a time. Short lessons, with frequent brief intervals in which the attention is permitted to relax, should be given at first.

12. The Effects of Alcohol on the Nervous System and their Symptoms.—Alcohol is a terribly frequent cause of nervous diseases. In over-stimulating the brain and spinal cord, it impairs their structure, weakens their functions, and often leads to insanity and crime.

A small quantity of wine or spirits, taken by one not accustomed to it, congests and excites the brain; the person gets restless and talkative, then dizzy and unable to think clearly. He is soon overcome by sleep, and on awaking feels out of sorts.

If the dose be increased, the talkativeness is accompa-

12. What is the action of alcohol on the brain and spinal cord? What is the action of a glass of wine on a person not used to it?

---

* Professor Oliver Wendell Holmes.

nied by indistinct speech and the dizziness by trembling hands and a staggering walk, both showing loss of control over the voluntary muscles and the will. The sense of touch is dulled; the eyeballs do not move together, so as to look exactly at the same point at the same moment, and objects, accordingly, appear double. (You may imitate this effect by pushing one eyeball gently while looking with both eyes at something.) Then follows profound drunken sleep, which may pass into " coma," a condition of deep unconsciousness from which the person cannot be aroused, and in which the breathing is slow and labored because the involuntary nerve-centres which govern the breathing-muscles are affected. Sometimes these centres become at last quite paralyzed and death results, but more often the man sleeps off his drunken fit, to awaken with a state of his nerves to be relieved only by renewed drinking, followed each time by worse results.

The nerve-centres, however, soon get used to the stimulant; it takes a larger amount each time to make them unsteady, but all the while brain and spinal cord are becoming surely, if slowly, diseased.

13. Some of the Nervous Diseases due to Alcohol.—*Delirium tremens* (trembling madness) is a frightful form of temporary madness, accompanied by great trembling. The senses are partly lost; the man sees spectres, usually foul and horrible, about him, and has all sorts of terrifying visions. He is at times violently excited and raving

What if the amount be increased? What is coma? Why is the breathing labored during coma? What may result? Why is one fit of drinking likely to lead to another? Why does it need more alcohol to make a practised toper drunk?

13. What is delirium tremens? Its symptoms? Its causes? Dip-

mad; in the intervals, utterly prostrate, sleepless, and a prey to indescribable terrors of the imagination.

Repeated drunkenness usually ends in an attack of this disease, but it is more frequently the result of continued hard drinking in persons who have never become actually drunk. It is especially apt to occur in those who drink to "keep them up" while engaged in hard mental work.

Dipsomania is a diseased condition, often only showing itself at long intervals, and marked by a mad passion for alcohol. However disgusting a liquid containing alcohol may be, the dipsomaniac will swallow it greedily. While the fit is on him he is as irresponsible as a madman, and his only safety is in being restrained as one.

This disease is sometimes produced by indulgence in drink, but is more often inherited from parents who have been drunkards. Sufferers from it are entitled to sympathy to which the common drunkard has no claim.

Paralysis, epilepsy, and *insanity* often result from drinking. There is, in fact, no kind of madness or of nervous disease which may not be, and has not been over and over again, produced by alcoholic drinks. Many of these diseases have other causes also, but none so frequent as alcohol.

Perhaps the greatest evil of intemperance is that the drunkard so often transmits to his innocent children some form of nervous disease. In the families of such are found the weak in body, weak in mind, weak in will,

somania? Symptoms? Treatment? Cause? Name other nervous diseases produced by drinking What is said of the causes of madness and nervous diseases? Of the transmission of such diseases to a drunkard's children? What do we find in the families of drunkards?

weak in character: the epileptic, the rickety child, the half-witted, the idiot, the dipsomaniac, the maniac; children who grow up unable to honestly make their way in the world, and become public burdens in insane asylums, prisons, or poorhouses.

CHAPTER XX.

NARCOTICS.

1. Narcotics.—Certain drugs have the power of making the cerebrum unable to work for a time; they thus cause unconsciousness, and produce what seems to be sound sleep. Substances which act on the nervous system in this way, are named *narcotics*. In small doses, they often relieve pain without causing actual loss of consciousness. Chloroform, chloral, ether, opium, laudanum, and morphia are examples of narcotics. Tobacco may be included, since, when not taken as a mere idle luxury, it is employed to soothe the nerves. Alcohol in large doses is also a narcotic. Occasionally, in a crisis of disease, when sleep must be obtained at any cost, or terrible pain is wearing out the strength of the sufferer, a narcotic, carefully ordered in proper dose by a physician, is a very valuable medicine. Taken habitually, narcotics weaken the mind, injure the whole nervous system, and cause many diseases.

2. Opium and Morphia.—Opium is a gummy mixture obtained from a kind of poppy Its chief active principle is *morphia*. The forms in which opiates are most used are: (1) *gum opium*, the natural substance, often put

1. What power have narcotics? What if taken in small dose? Give examples of narcotics. When is tobacco one? Alcohol?
2. What is opium? Morphia? What are the commonest forms of opiates?

up in the form of pills; (2) *laudanum*, made by dissolving opium in alcohol; (3) *paregoric*, a liquid containing several ingredients, of which opium is the most important; (4) *morphia*, and solutions containing it.

3. The Opium Habit.—Opium is perhaps the most valuable drug at the disposal of the physician. On the other hand, it is one of the most hurtful substances used by mankind. It may be that it does not do as much harm in the United States as alcoholic drinks, but only because not so many persons have taught themselves to crave it. Used constantly, it is as surely fatal, and the habit is perhaps even harder to break, for it may be indulged more secretly, and its effects are not so readily recognized. There is this, also, to be said: most of those who kill themselves by drink are persons of weak will, while many a one of highest gifts and noblest character, who would loathe the low vice of drunkenness, has, before knowing the danger, become the hopeless victim of opium. Using the drug, at first, as ordered by a physician for the relief of pain, he (or· she, for more women than men are given to opium-excess) is scarcely conscious of danger, until the repeated employment of the drug has created an almost irresistible craving for its continuance. Most medical men now fully recognize the danger, and only order prolonged use of opium with great caution.

4. The Diseased Conditions Produced by Regular Use of Opium.—The first effect is deadening of sensibility, accompanied by mental exaltation, if the dose be small.

3 What is said of opium? Of its harmfulness as compared with alcohol? Why is opium more disastrous from one point of view? How is it now given by physicians?
4. What are the first effects of a dose of opium? What is the con-

This is succeeded by unnatural sleep, disturbed by fan-tastic dreams.

On awaking, there is great depression of mind and body: often associated with defective memory, and a feeling that something terrible is about to happen. There is muscular weakness; distaste for food, without actual nausea; and an almost irresistible craving for an-other dose.

If the habit be continued further, mental and physical changes occur. Distaste and inaptitude for any kind of exertion; weakened digestion; not enough secretion of bile; slow action of the muscles of the bowels, causing constipation. The voluntary muscles waste, the skin shrivels, and the person gets the appearance of old age prematurely. The pulse is quick, the body feverish; the eye dull, except just after taking a dose of the drug.

Next comes failure of the nervous system. The legs are partly paralyzed, and then the muscles of the back. The victim crawls along, bent like an old man. Death finally results from starvation, due to complete failure in the working of the digestive organs.

5. Morphia or Morphine.— When morphia is used, a solution of it is usually injected under the skin by a sharp-pointed syringe. Continued use of it in this or any other way is followed by all the symptoms of opium-poisoning above described, and has the same fatal ending. The digestive organs are not so quickly injured; but, on the other hand, the repeated punctures of the skin cause inflammation and sores.

dition of the person on awaking? What results follow continuance of the habit? How does opium affect the nervous system?

5. How is morphia usually given? Results of its continued use? Compare its effects with those of opium.

6. Danger of Administering Opiates to Children. —
Children are extremely easily poisoned by opium and all
things containing it or morphia. *They should never be
given to a child except on the order of a physician, and exactly
as ordered.* Many an infant has been killed by paregoric
or some "soothing syrup" containing opium, given, without medical advice, by a parent or nurse to stop diarrhœa or produce sleep.

7. Chloral, Chloral Hydrate, Syrup of Chloral. —A few
years ago, chloral was proclaimed a wonderfully harmless narcotic: it caused sleep or lessened pain without
harm, it was said, to mind or body. · Physicians have
since learned that it is not at all the harmless drug they
formerly believed it, but many other people have not yet
had their eyes opened to its dangerous character. Various preparations containing it are sold in drugstores to
any one asking for them; and many persons who would
hesitate to take opiates without medical advice, use
chloral, believing it quite safe and harmless.

Chloral, taken habitually, is at least as mischievous as
opium. To retail it in any form except on the prescription of a physician, should be made illegal.

The chloral habit is acquired with great ease, and is
very hard to break. The first phenomena of chloral disease (*chloralism*) are these: The digestion is greatly impaired; the tongue is dry and furred; there is nausea;
sometimes vomiting, and a constant feeling of oppression from wind on the stomach.

6. Why should opiates never be given to a child except by a physician's order? What has resulted from neglect of this precaution?

7 What was believed of chloral a few years ago? What have
medical men lately learned about it? Why do so many people take
chloral without medical advice? Describe the first symptoms of

Next, nervous and circulatory disturbances occur. The temper becomes irritable, the will weak; the hands and legs tremulous; the heart-beat irregular; the face easily flushed. Sleep becomes impossible without use of the drug, and when obtained is troubled, and the person awakes unrested.

In later stages, the blood is seriously altered. Its coloring matter is dissolved out of the corpuscles into the plasma (p. 135), and then soaks through the walls of the capillary vessels, causing purplish patches on the skin.

If the chloral-taking be still continued, death results from impovished'blood, weakened heart, or paralysis of the nervous system. Not unfrequently, chloral-takers unintentionally commit suicide by indulging in too large doses.

8. Bromides.—The drugs included under this name, resemble chloral and its compounds in that they were once regarded as safe soothers of the nervous system and promoters of sleep, that physicians have now learned that they are very dangerous when frequently used, and that the general public still believe them safe, and often use them without a doctor's advice. They are very valuable medicines in some circumstances, but may do nearly as much harm, when taken indiscreetly, as opium or chloral. Some mothers and nurses who have learned the danger of paregoric and soothing syrups, now give bromides instead to restless infants. The bromide may not be so dangerous as the opiate, but it should never be given except on a doctor's prescription.

chloralism. What are the symptoms in more advanced chloralism? What in the latest stages?

8. In what do bromides resemble chloral? What are the dangers of using them? What precautions are necessary?

9. Tobacco is often indulged in for the sake of soothing the nervous system or lessening the feeling of mental fatigue or worry. It also decreases the oxidations of the body, and its wasting, and so enables it to get along with less food; it may in this way be useful to a starving or ill-fed person. It contains a small amount of an active principle, *nicotin*, which is a powerful poison. A few drops of pure nicotin will cause rapid death by paralyzing the heart. When tobacco is smoked, some of the nicotin is burned; but vapors containing ammonia are formed, and these irritate the mouth and throat. The ill effects of smoking are thus, in part, general—due to absorbed nicotin; and in part local—due to irritating matters in the smoke. It cannot be denied that many persons consume a good deal of tobacco without being much harmed by it. But it does no one any good unless he cannot get sufficient food, or his nervous system is so diseased or irritable that it needs soothing. One general rule may be laid down without fear of contradiction: *tobacco is always very injurious to those whose bodies are not yet fully developed.*

10. The Local Action of Tobacco is at first manifested by an increased flow of saliva. After some practice in smoking this effect ceases, and is succeeded by a feeling of dryness in the mouth, which often leads to indulgence in alcoholic drinks. In this perhaps lies the greatest danger from tobacco. The habitual smoker often suffers from what is well known to physicians as

9. Why is tobacco indulged in ? When may its use be beneficial ? What is said of nicotin ? What becomes of it when tobacco is smoked ? The ill effects of smoking ? What general rule may be safely stated ?
10 How does the local action of tobacco first show itself ? How is this changed by practice in smoking ? Point out one of the chief

"smoker's sore throat." This is accompanied by a hack-
ing cough, and often with difficulty in speaking and some
deafness. Cure is impossible unless smoking is given
up.

The smoke of the paper in which cigarettes are rolled
especially irritates the throat and larynx. So far as
these organs are concerned, a cigarette is the most injurious form in which tobacco can be smoked.

11. The General Action of Tobacco.—The absorption of
nicotin and other substances contained in tobacco, is apt
to interfere with the proper development of the red corpuscles of the blood. This, as you have learned (p. 137),
is a very serious evil, because these corpuscles have to
carry oxygen all through the body for use by the different organs. As a result of their deficient quantity, not
only does the skin grow pale, but all the organs do poor
work. The muscles become feeble; the stomach digests
badly; the heart is weakened and subject to attacks of
palpitation; and the eyesight very often impaired. In
general, there is produced a feeling of lassitude and in-
disposition to exertion of any kind that, in view of the
heavy odds a man has to contend against in the struggle
of life, may prove the handicap that causes his failure.
If success in life be an aim worth striving for, it is surely
unwise to shackle one's self with a habit which cannot
promote and may seriously jeopardize it.

dangers from tobacco. What is smoker's sore throat? By what accompanied? What necessary for cure? What is said of cigarettes?
11. Action of absorbed nicotin on the blood? Why serious? Action of nicotin on the muscles? The stomach? The heart? The
eyesight? What is said of its effects in general?

CHAPTER XXI.

THE SENSES.

1. Common Sensation and Special Senses.—Each of us has a great many *feelings*, or *sensations*, of different kinds. We may be hungry or thirsty or tired or suffer pain in a variety of ways. Such sensations as these tell us about our own bodies. Hunger warns us to eat, nausea or "sickness" that the stomach is not in a condition to digest, pain that some part is diseased or injured and needs attention. All these kinds of feeling are named *common sensations.*

Other kinds of sensations enable us to learn about things outside of our bodies, and to perceive and use objects in the world around us. These sensations are known as the "special senses;" they include sight, hearing, smell, taste, and touch, which are commonly spoken of as "the five senses." To these we should add the *temperature-sense*, which often enables us to learn that something is hot or cold without touching it or seeing it. These senses have been well called the "gateways of knowledge," because without them the mind would have to remain in complete ignorance of the world and universe in which we live.

1. How do we learn the needs of our own bodies? Examples? What are these feelings called? What is the use of the special senses? Name the "five senses." What is the temperature sense? Why are the senses called the gateways of knowledge?

2. All Kinds of Sensation Depend on the Brain.—You have already learned that when the neive-fibres of the foot are cut anywhere on their way to the brain, the foot loses feeling (p. 199). This is true of every other part of the body which has feeling, whether it be merely a part possessing some common sensation, or an organ of one of the special senses. Also, if the brain be acted upon by chloroform or ether, or certain parts of it be seriously diseased or injured, feeling is lost, although the nerves and the sense-organs may be quite unaffected. We thus learn that all feeling is due to some change in the brain. Usually, when we have a sensation, whether of sight, hearing, pain, or any other kind, it is due to the fact that some sensory nerve (p. 196) has been set at work, and has carried a message to the brain. This message has then set at work, or excited, a part of the brain, which makes us see or smell; and so on. The mind has learned from what parts of the body these messages to the brain usually start, and we have come to think of each kind of feeling as being in the organ or place from which the message starts, and not in the brain itself. When the eye is closed we do not see, so we think the sense of sight is in the eye. Yet it really is in the brain: all that the eye does is, when light acts on it, to send messages along its nerve to the brain, and set to work that part of the brain which has feelings of sight; and so it is with our other senses.

2. What results when all the nerve-fibres coming from any part of the body which has feeling, are cut? How may loss of feeling be caused without affecting the sense-organs or the nerves? What is thus proved? To what is a sensation usually due? What happens when the message sent by the sensory nerve reaches the brain? How has the mind come to connect certain feelings with certain parts of the body? Illustrate. What sometimes happens as regards sensations in disease? Results?

Sometimes in disease, the parts of the brain which give us feelings are excited without waiting for any message brought in along a sensory nerve. Then the person becomes delirious or suffers from delusions. He sees and hears and smells things which do not really exist; but to his mind they are just as real as if they did actually exist, and were acting on his sensory nerves so as to excite the parts of the brain which feel.

3. **Why the Eye is the Organ of Sight.**—In the eyes there are thousands of nerve-fibres, each of which has a little "tip" or end on it which is so made as to be very easily acted on by light. Any light, as from the sun or a lamp or candle, which comes direct, or is first reflected from some object, and reaches one of these peculiar little ends, excites it, and the end in turn excites the nerve-fibre joined to it, and this fibre then carries up some message to the part of the brain which gives us feelings or sensations of sight. If the light comes direct it excites the nerves in such a way that we see the sun or lamp or candle. If it comes bounding back from some other object which it has struck on its way, we see the other object. No other nerves than those of the eye have this particular kind of tips on their ends, and so light does not excite them, as it does the nerves of the eye.

4. **The Eyeball** (Fig. 55) is nearly as round as a marble, but is buried in the eye-socket and covered by the eyelids, so that only a small part of its front side can be seen. On this front part is a round transparent window, set in it, like a pane of glass, to allow light to get

3. What is there on the ends of the nerve-fibres in the eye? What happens when light reaches them? Why cannot we use other parts of the body for seeing?
4. Describe the shape and position of the eyeball. What is there

into it. To the inner or deeper side of the eyeball is attached the *optic nerve,* 17, which runs to the brain, and is the nerve of sight.

5. The Eyeball has Three Coats, an outer, a middle, and an inner.

The outer coat is tough and strong: on the back and

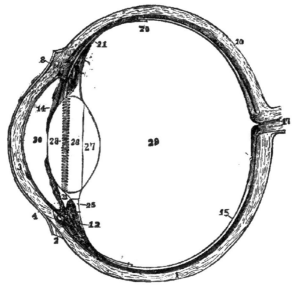

FIG. 55.—The left eyeball in horizontal section. 1, sclerotic; 2, junction of sclerotic and cornea; 3, cornea; 10, choroid; 14, iris; 15, retina; 17, optic nerve; 26, 27, 28, are placed on the lens; 29, vitreous humor; 30, aqueous humor.

sides of the eyeball it is opaque, that is to say, does not let light go through it. A little of it can be seen between the eyelids, as the " white of the eye." The opaque part of the outer coat is named the *sclerotic* (1, Fig. 55).

on its front? Where does the optic nerve join it? Where does the nerve go? Function of this nerve?

5. What coats has the eyeball? Nature of the outer. Describe the sclerotic. The cornea. The choroid. The iris. The pupil. Why

The front part of the outer coat is the transparent portion above mentioned. It is called the *cornea* (3).

The middle coat is colored. Its hinder portion, 10, is black, and lies close against the sclerotic; it is called the *choroid*. Its front part separates from the outer coat, and instead of lying close against the cornea, turns in a little way behind it, 14, so as to leave a space, 30, between. This part of the middle coat is called the *iris*. Its color varies; we see it through the cornea, and say the eye is brown, or blue, or gray, or black, according to the color of the iris. In the middle of the iris is a hole, *the pupil* of the eye. It looks black, just as a hole opening into a box whose inside was painted black would, if you viewed it from outside, although the hole would let light into the box. The dark choroid answers to the black paint inside the box; in some animals, as dogs and cats, part of it is not black, and so the inside of the eyes of those animals, seen through the pupil, often looks shining. In bright light, the pupil becomes smaller, so as to protect the nerves inside the eye from being over-stimulated and dazzled: when there is not much light the pupil becomes larger. If you stand in front of a mirror and close your eyes for half a minute, and then open them and let light get into them, you can watch your pupils getting smaller.

The inside coat of the eyeball is the *retina*, 15. It is very thin, and is transparent so that the dark color of the choroid shows through it. The retina only lines the hinder half of the eyeball. It is the sensitive part of

does the pupil look black? What is said of its expansion and contraction? How can you see the contraction of your own pupil? What is the retina? Describe it. Its position? Of what does it consist? Illustrate the connection of optic nerve and retina.

the eye, and consists of the spread-out fibres of the optic nerve, and the peculiar tips or "end organs" joined to them. If you should take a cord, and fray out its threads at one end, and spread them out on all sides, the cord would answer to the optic nerve, and the spread-out threads to its fibres in the retina, except that each thread, in order to make the resemblance greater, ought to have a very small rod or cone easily excited by light, attached to its end.

6. **The Interior of the Eyeball** is filled up by liquid or jelly-like matters, surrounded by its coats, as the pulp of an orange is surrounded by the rind. These substances are all transparent; they guide to the retina, light which enters the eye through the cornea and pupil. They are three in number. (1) The crystalline lens, 26, 27, 28, just behind the iris. It is soft and jelly-like. (2) The aqueous (watery) humor, 30, a watery liquid between the crystalline lens and the inner side of the cornea. (3) The vitreous (glassy) humor, 29, behind the crystalline lens, a soft jelly filling up all the back part of the cavity of the eyeball.

7. **The Use of Aqueous Humor, Lens, and Vitreous Humor** is to gather the rays or lines of light which enter the eye, and so bend and direct them, that all those starting from one point outside the eye meet again in one point on the retina, and excite the same nerve-fibre. This enables us to see things distinctly, because an exact image of the thing looked at is made on the retina. In Fig. 56, *O* answers to the lens of the eye; *D, E,* is the object looked

6. How is the interior of the eyeball filled? Use of these substances? Their number? Names? Describe each.
7. What is the use of aqueous humor, vitreous humor, and lens? How does their action enable us to see distinctly? How is the image

at; and *d, e,* its image on the retina. The image is much smaller than the object, and is wrong side up, but the mind has learned by experience to understand it in the right way.

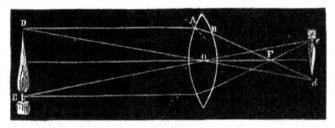

Fig. 56 —Illustrating the formation behind a convex lens of a diminished and inverted image of an object placed in front of it.

8. Short-Sight and Long-Sight.—When you use a telescope or an opera-glass to look at any object, you have to focus it. The arrangement which will enable you to use it for seeing near objects distinctly, must be changed before you can use the glass for seeing things farther off. In our eyes, the lens does this focusing; it changes according as we look at near or distant things. In persons with good eyes (*A*, Fig. 57), the lens can accurately focus on the retina, images of very distant objects, and also of things within seven or eight inches of the eye. In other persons (*B*), the eyeball is too long from front to back, and the lens cannot focus on the retina the rays or lines of light coming from distant objects: such persons are *short-sighted*. They can see very distinctly things near the eye, but more distant objects seem

of an object looked at, depicted on the retina? Why do we see it rightly?

8. How is a telescope arranged for seeing near or distant objects? How do our eyes focus what they look at? What is said of this power in good eyes? Why are some eyes short-sighted? Why are others long-sighted?

blurred and indistinct. The opposite defect is *long-sight*. In those who suffer from it, the eyeball is so flat that the lens cannot focus on the retina rays of light coming from a near object (*C*, Fig. 57).

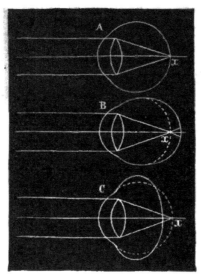

FIG. 57.—Diagram illustrating the path of parallel rays of light after entering a healthy, well-shaped eye (*A*), a short-sighted eye (*B*), and a long-sighted eye (*C*).

9. Hygiene of the Eyes.—Looking directly at very bright objects, as the sun or an electric lamp, dazzles and injures the eyes; so does sudden change from darkness to light. On first waking, the eyes should be somewhat gradually accustomed to bright light. The ill effects of such changes are much less serious than the harm that may be done by using the eyes when there is not enough light to see clearly. Frequent reading or

9. What is the effect on the eyes of looking at very bright lights? Why should they be gradually accustomed to light after sleeping? What is even more injurious to the eyes than sudden changes from

sewing in such feeble light that the eyes feel strained, will certainly injure them permanently.

Long-sight and short-sight are not diseases. They are due to the fact that the eyeball is not perfectly shaped, but it may, nevertheless, be perfectly healthy. Both defects are easily remedied by proper spectacles or eye-glasses. If neglected, they lead not only to disease of the eye itself, but to headaches, and other symptoms of nervous disorder.

10. The Eyelids are folds of skin moved by muscles so as to cover or uncover the front of the eyeball, or, as we ordinarily say, to shut or open the eye. Opening along the edge of each eyelid, are twenty or thirty small glands. Their secretion is greasy and keeps the tears from flowing over the edge of the eyelids, except when they are secreted in large quantity. The eyelid-secretion is sometimes too abundant, and then appears as a yellowish matter along the edges of the eyelid. It often dries during the night and causes the lids to be glued together in the morning.

11. Tears are secreted by the *tear* or *lachrymal* glands, which lie, one in each eye-socket, above and to the outer side of the eyeball. They are poured on the front of the eye by the tear-ducts which open on the deeper or inner side of the upper eyelid, near its outer corner. Tears are secreted all the time, but usually only in small quantity. Winking spreads them all over the front of

darkness to bright light? How may short sight or long sight be remedied? What happens if they are neglected?

10 What are the eyelids? What open along their edges? Use of these glands? Why are the eyelids sometimes glued together in the morning?

11. Where are the tear-glands? Where do these ducts open? How

the eyeball, and they keep it moist. What remains is drained off by canals which run from the inner corner of each eyelid to the inside of the nose, from which the liquid flows into the pharynx, and is swallowed. In weeping, the tears are secreted faster than these canals

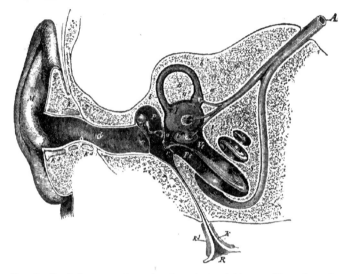

FIG. 58.—Semi-diagrammatic section through the right ear. *M*, concha. *G*, external auditory meatus. *T*, tympanic or drum membrane. *P*, Tympanum. Extending from *T* to *o* is seen the chain of tympanic bones. *R*, Eustachian tube. *V, B, S*, bony labyrinth; *V*, vestibule; *B*, semicircular canal; *S*, cochlea. *b, l, l'*, membranous semicircular canal and vestibule. *A*, auditory nerve dividing into branches for vestibule, semicircular canal, and cochlea.

can carry them off, so they flow over the lower eyelids and trickle down the face.

12. Hearing.—The ear consists of three portions, known as the *external ear*, the *middle ear* or *tympanum* (drum), and the *internal ear* or *labyrinth*. The labyrinth is so named

are they spread over the eye? Where are they usually carried from the eye? Why do they trickle down the face in weeping?
12. What are the three main portions of the ear? Why is the

because it has many winding passages in it. The nerves of hearing are the two *auditory nerves.* One runs to each ear from the brain, and its fibres end in the labyrinth, in connection with peculiar very small organs which are easily excited by slight shaking, and then excite the fibres of the auditory nerve. Everything that gives out sound shakes or vibrates, and sets the air all round it shaking. The use of the outer ear and middle ear is to take up the vibrations of the air and pass them on to the organs on the ends of the nerve-fibres in the inner ear.

13. The External Ear consists of the expansion (*M*, Fig. 58) seen on the exterior of the head, called the *concha* (shell), and a passage leading in from it, the *external auditory meatus, G.* This passage is closed at its inner end by the *tympanic*, or *drum, membrane, T.* It is lined by a prolongation of the skin, through which numerous small glands, secreting the *wax* of the ear. open.

14. The Tympanum, or drum-chamber of the ear (Fig. 59 and *P*, Fig. 58), is a small cavity in one of the bones on the side of the skull. It is closed externally by the drum-membrane From its inner side the *Eustachian tube* (*R*, Fig. 58) proceeds and opens into the pharynx (*g*, Fig. 30). This tube allows air from the throat to enter the tympanum, and serves to keep equal the pressure of the air on each side of the drum-membrane. Three small bones (Fig. 59) stretch across the tympanic cavity from the drum-membrane to the labyrinth; they

labyrinth so named? What are the auditory nerves? What is attached to ends of their fibres in the ear? How used in helping us to hear? Use of outer and middle ear?

13 Of what does the external ear consist? What is found at the inner end of its passage? How is the passage lined?

14. Describe the tympanum. What is the Eustachian tube? Its

pass on to the labyrinth, the vibrations of the membrane, produced by vibrations of the air. The outmost bone is the *malleus* or *hammer-bone, L ;* the inmost, the *stapes* or *stirrup-bone, S ;* and the one between, the *incus* or *anvil-bone, H.*

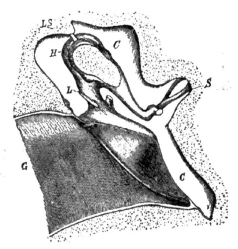

Fig. 59.—The tympanic cavity, *C, C,* and its bones, considerably magnified. *G,* the inner end of the external auditory meatus, closed internally by the conical tympanic membrane; *L,* the malleus, or hammer-bone; *H,* the incus, or anvil-bone; *S,* the stapes, or stirrup-bone.

15. The Internal Ear, or Labyrinth, consists of chambers and tubes hollowed out in the inner part of the temporal bone, *T,* Fig. 6, and containing thin bags and tubes, filled and surrounded by watery liquid. Inside these bags and tubes the fibres of the auditory nerve end. Its middle chamber, called the *vestibule* (*V,* Fig. 58), has an opening, the *oval foramen, o,* in its outer side, into

use ? Number and arrangement of the bones in the tympanum? Their use ? Names and position of these bones ?

15. Of what does the internal ear consist ? Where do the fibres of the auditory nerve end ? Name of the middle chamber of the internal ear ? Where is the oval foramen, and what fits into it ? Where are

which the inner end·of the stapes, or stirrup-bone, fits.
Behind, the vestibule opens into three *semicircular canals*,
one of which is shown at *B*, and in front into a spirally
coiled tube, *S*, the *cochlea*.

When shakings or vibrations of the air make the tympanic membrane vibrate, it shakes the tympanic bones;
the stapes, vibrating in the oval foramen, then shakes
the liquids in the labyrinth, and sets up vibrations in
them, which excite the endings of the auditory nerve.
The stimulated auditory nerve then conveys a nervous
impulse to the part of the brain concerned with hearing
and excites it, and a sensation of sound results.

16. Hygiene of the Ear.—The outer parts of the ear
are less tender than the eye, and are more often injured
by unnecessary meddling. When the ear is healthy, its
wax dries up into scales and is shed in proper quantity.
Some of it is necessary to protect the inner parts of the
ear. Rubbing it out by stiff objects, not only removes
it too fast, but may cause inflammation of the tympanic
membrane. If the wax is clearly excessive, or if there is
any running from the ear, it is wisest to consult a physician at once. No stiff rod should ever be put into the
ear, except by a skilled person. The tympanic membrane is very thin and may easily be torn. Young children often put such things as peas and small beans in
their ears. If they do not come out very easily, get a
doctor to remove them. In any such case, do not pour

the semicircular canals? The cochlea? Describe how the endings of
the auditory nerve are excited by vibrations of the air. What results
when the auditory nerve is stimulated?

16. Why are the ears more often injured than the eyes by meddling?
What happens to the wax of the ear in health? Why is some wax
necessary? What may result from removing it? What should be
done when there is any running from the ear? When a child has put

water into the ear; it causes a pea or bean to swell, and makes its removal very difficult.

Deafness may be caused in many ways: by disease of the auditory nerve, by disease of the labyrinth or of the tympanum, by stoppage of the outer passage by wax or some foreign object, or by inflammation and swelling of the membrane lining the Eustachian tubes. Swollen tonsils (p. 101). or a cold which has settled on the throat, or smoking, very often cause deafness in the way last mentioned. If the auditory nerve or the internal ear are at fault, the deafness may be incurable In most other cases, cure is possible with medical aid. In the case of a cold, the cure usually occurs of itself if you have a little patience.

17. Touch, or the Pressure-Sense.—Many sensory nerves end in the skin, and through it we get several kinds of sensation; *touch, heat* and *cold*, and *pain ;* and we can with more or less accuracy say from what parts of the skin they have come. The interior of the mouth also possesses these feelings. Through touch, we recognize pressure on the skin, and the force of the pressure; the softness or hardness, roughness or smoothness, of the body producing it; and the form of this body, when it is not too large to be felt all over. The nerves of touch are very numerous. A great many of them end inside papillæ of the dermis (p. 63).

18. The Delicacy of the Sense of Touch is very different

some foreign body into its ear? Name some of the causes of deafness. How may swollen tonsils cause deafness? When is deafness apt to be incurable?

17 What sensations do we get from the skin? What other part of the body gives rise to these sensations? What do we recognize through touch? Where do many of the nerves of touch end?

on different parts of the skin. It includes two distinct things, which are often confounded. In the strict sense of the words, touch is most delicate where the smallest pressure can be felt. In this meaning, the sense of touch is most acute on the forehead and temples, where a lighter weight can be felt than on any other part of the skin. Usually, however, by delicacy of touch is meant the accuracy with which, the eyes being closed, we can tell the exact point of the skin which is touched. In this meaning, the sense of touch is most acute on the tip of the tongue, the edge of the lips, and the ends of the fingers. If the blunted points of a pair of compasses, closed to within one twelfth of an inch, be gently laid, at the same moment, on a finger-tip, we distinguish between them and feel two touches, while on the back of the neck they must be more than an inch apart before we can distinguish them. The papillæ of the dermis are always numerous where the distinguishing power is great.

19. The Temperature-Sense.—By this is meant our faculty of perceiving cold and heat; and, with the help of these sensations, of perceiving whether things are cold or hot. Its organs are the whole skin, the mucous membrane of mouth, pharynx, and gullet, and of the entry of the nose. Burning the skin will cause pain, but not a true temperature-sensation, which is quite as different from pain as touch is.

18. What is meant by delicacy of touch in the strict sense of the words? Where is it most acute? What is usually meant by delicacy of touch? Where is it most acute? Give an illustration of its variation on different regions of the skin. Where are the papillæ numerous?
19. What is the temperature-sense? What are its organs?

20. Smell.—The organ of smell, or the olfactory organ, consists of the mucous membrane lining the upper portions of the two nostril-cavities. Part of it is shown at *o* and *p*, Fig. 42. The nerves of smell are the two olfactory nerves, one of which runs from each nostril-chamber to the brain.

21. Odorous Substances frequently act powerfully when present in very small quantity. A grain or two of musk kept in a room will give the air in it an odor for years, and yet at the end will hardly have diminished in weight, so infinitesimal is the quantity given off from it to the air and able to excite the sense of smell.

22. Taste.—The organ of taste is the mucous membrane on the upper side of the tongue, and the under side of the soft palate (p. 101), The mucous membrane of the tongue presents innumerable elevations or papillæ. Some are organs of touch, for the tongue has the sense of touch as well as of taste. Others contain the endings of nerve-fibres which, when excited, stimulate the taste-centres in the brain and cause sensations of taste.

Many so-called tastes (flavors)· are really smells ; particles of substances which are being eaten reach the nose through the pharynx (see Fig. 42), and arouse smell-sensations which, because they accompany the presence of objects in the mouth, we take for tastes. Such is the case with most spices; when the nasal chambers are blocked during a cold in the head (p. 154), or closed by holding

20. Of what does the olfactory organ consist?
21. Illustrate the efficiency, so far as producing smell-sensations is concerned, of a very small quantity of an odorous substance.
22. What is the organ of taste? What is found on the mucous membrane of the tongue? What are the uses of its papillæ?
What are many so called tastes? Illustrate.

the nose, the so-called "taste" of spices is not perceived when they are eaten. If cinnamon, e.g., is chewed under such circumstances, the only sensation felt is a sort of hot

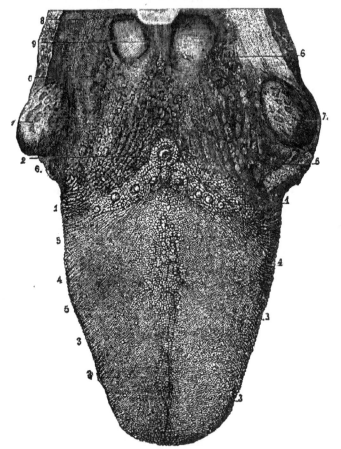

FIG. 60.—The upper surface of the tongue. 1, 2, circumvallate papillæ; 3, fungiform papillæ; 4, filiform papillæ.

feeling in the mouth. Some of the most nauseous medicines have really no taste, or very little. If the nose be held, they can be swallowed without disgust.

CHAPTER XXII.

THE ACTION OF ALCOHOL ON BODY, MIND, AND CHARACTER.

1. Introductory.—We hope that the boys and girls for whom this book has been written, with its statement of the structure and working of the parts of the human body, and the rules which must be observed if health is to be kept, have had little chance to gain experience of the evils of intemperance. Unhappily, none of us can remain long ignorant of them. All around us are those who suffer in one way or another from the effects of alcoholic drinks. We speak not only of those who themselves indulge in them, but of the far larger number whose lives are spoiled by the ruin of their natural protectors and their loved ones.

We do not mean to say that most of those who drink liquor are drunkards, or indulge in it to excess, in the ordinary sense of the words; but when we think of the great number who daily take drinks containing alcohol; when we call to mind the fact that what is usually called moderate drinking, which never makes a man drunk, is often positively hurtful, and may alter for the worse nearly every important organ of the body; when we re-

1. Why are we unlikely to remain ignorant of the evils of alcohol-drinking? What is said of "moderate" drinking? Of nervous diseases due to alcohol? Of its general effect on human happiness?

member that nervous diseases are very frequently produced by alcohol, and are more often transmitted by parents to their children than any other class of diseases, assuming worse forms as they are passed on from generation to generation; when we recall such facts, we have no reason to wonder that more disease and premature death, more crime and misery, are due to alcohol than to bad drainage, foul air, insufficient food, unsuitable clothing, or any other of the subjects treated of in an elementary text-book of physiology and hygiene.

The habit of drinking is often formed in ignorance of its consequences. Be warned and instructed in time, to protect yourself and others against it. Many of the diseases produced by alcohol come on so gradually that they are not recognized until the will has become too weak to resist what the appetite craves.

The form of disease depends on the sort of drink, the amount, and the constitution. Some few there are, whose excretory organs are so active that the alcohol is quickly passed out of the body, and no disease, due to it, manifests itself until the close of, perhaps, a long life. Such persons are, however, marked exceptions to the general rule, which may be thus stated: *prolonged excessive use of alcoholic liquors, leads surely to disease of the body and disease of the mind; often to insanity and death.*

2 Alcoholic Drinks, as you have already learned (p. 94), are all such intoxicating liquors as brandy, whiskey, gin. rum, wines, ales, beer, and cider ; also mixtures which

How is the habit often formed? Why are the diseases caused by it often discovered too late ? On what does the form of disease depend ? Why do some persons escape for a long time ? What is the general rule ?

2. Name the alcoholic drinks most often used.

contain them, as cordials, punch, egg-nogs, and many " tonics."

We have studied their effects upon some of the most important organs of the body in turn; but in order to fix them more clearly in our minds, let us review the whole subject.

3. Alcohol as a Food —Foods build tissues; alcohol leads to overgrowth of some tissues, but not to growth of muscle, brain, or gland. Foods supply strength or working power; alcohol stimulates brain and muscle to overwork, and as it nourishes neither, the final result is failure in strength and endurance. Foods maintain animal heat; alcohol makes one feel warm for the moment, but its actual effect on the temperature of the body is to lower it (pp. 96, 97).

4. Effects of Continued Use of Alcohol on Various Tissues and Organs.—These may be summed up as follows:

Connective tissue is so increased in quantity that it crushes and destroys parts which, when present in only healthy amount, it protects (p. 14).

The muscles have their strength, and their power of keeping a long time at work, lessened (p. 58). They are also made liable to chronic rheumatism (p. 50).

The skin has its vessels dilated and an excessive amount of blood made to flow to it, causing congestion; and impairing that activity of its glands necessary to maintain health (p. 76).

The digestive organs in general are often diseased in

---

3 What is said of alcohol as regards the building of tissues? As a strengthener and stimulant? In regard to its effect on the temperature of the body?

4. Action of alcohol on connective tissue? On the muscles? The

consequence of the general slow poisoning of the body caused by alcohol. The stomach and liver are more directly attacked by it.

1. The mucous membrane of the stomach becomes congested, then inflamed. It fails to secrete gastric juice and indigestion results (p. 130).

2. The true liver-substance being injured or destroyed by increased growth of connective tissue, the organ becomes a shrunken rough mass, unfit to perform its important duties in the nourishment of the body (p. 132).

The blood has its power of absorbing and carrying oxygen decreased, and also its power of clotting. Hence the temperature of the body and its working power are lessened, and any wound is more apt to bleed dangerously (p. 161).

The arteries have their walls weakened so that they become liable to burst under the pressure of the blood inside them (pp. 161, 162).

The heart has its beat quickened so that it does not get enough rest. Its overworked muscle thus does not get sufficient nourishment, and at last becomes unable to pump the blood along (p. 162).

The respiratory organs have their lining mucous membrane congested and irritated, increasing the liability to colds and other diseases (p. 184).

The kidneys are overstimulated, and at last become unable to do properly their work of removing nitrogen wastes. Very often a fatal malady, named Bright's disease, is produced (p. 189).

skin? The digestive organs in general? The stomach? The liver? The blood? The arteries? The heart? The respiratory organs?

The brain and spinal cord are kept in a chronic state of congestion *·and overexcitement. This results at first in inflammatory disease (delirium tremens); later in paralysis, epilepsy, or insanity (pp. 212, 214).

The senses are dulled, partly from disease of the nerves and nerve-centres, partly by diseased changes in the sense-organs.

No tippler probably ever suffered from all of the diseases above mentioned, and most of them may develop in persons who are total abstainers, but some of them are pretty sure to develop in habitual drinkers, and they are all more frequently due to intemperance than to any other single cause. It is also well known that in any serious disease, the chances of recovery are smaller in the case of drinkers.

5. Continued Alcoholic Indulgence Causes Premature Old Age.—Many of the alterations in various tissues and organs above described as brought about by alcohol, are very like the changes which naturally occur in old age. When alcohol does not cause some actual disease, as Bright's disease, or delirium tremens, it often hastens the ageing of the body. The organs lose strength and activity, and become old before their time.

The kidneys? The brain and spinal cord? The senses? How does habitual drinking affect the chances of recovery from disease?

5. What does alcohol often do when it does not ·cause actual disease?

---

* "I once had the unusual though unhappy opportunity of observing the same phenomenon in the brain-structure of a man who, in a fit of alcoholic excitement, decapitated himself under the wheel of a railway-carriage and whose brain was instantaneously evolved from the skull by the crash. The brain itself entire was before me within three minutes after death It exhaled the odor of spirit most distinctly, and its membranes and minute structures were vascular in the extreme. It looked as if it had been recently injected with vermilion."—Dr B W. Richardson.

6. The Destruction of Will and Character by Alcohol.—
One of the first effects produced by alcoholic drinks
is weakening of the control of the will over the actions.
A slightly tipsy man laughs and talks loudly, says and
does rash things, is enraged or delighted without due
cause. If the amount of alcohol be increased, the
power of the will is further lessened. The muscles
obey it very imperfectly, so speech becomes indistinct
and the legs unsteady. At the same time, the reason
is so weakened that the man is the prey of every tran-
sient whim: he is, by turns, affectionate and cruel, dar-
ing and craven, buoyed by hope and crushed by despair,
arrogant and full of shame, with no sufficient cause.

Habitual excessive use of alcohol thus soon leads to a
state in which the emotions are permanently overexcited,
and the will enfeebled. The man's highly emotional
state exposes him to special temptations, to excess of all
kinds of passion, and his weakened will decreases his
power of resistance. The final result is a degraded moral
condition. He who was prompt in the performance of
duty begins to shirk that which is irksome; energy gives
place to indifference, truthfulness to lying, integrity to
dishonesty, for even with the best intentions in making
promises or pledges, there is no strength of will to keep
them ; the man at last becomes regardless of every duty,
and even unable to accomplish any which momentary
shame may make him desire to perform.

6. Point out one of the first effects of alcoholic drinks. How illustrated ? If the amount is increased, what happens ? Illustrate
from the muscles ? How is the weakening of the reason in a drunken
man exhibited ? To what does habitual excessive use of alcohol lead ?
What are the consequences ? The final result ? What is the only
hope for an habitual drunkard ?

For such a one there is but one hope—confinement in an asylum where, if not too late, the diseased craving for drink may be gradually overcome, the prostrated will regain its ascendency, and the *man* at last gain the victory over the *brute*.

GLOSSARY.

Ab-dŏ'men (*Lat. abdere*, to conceal, *omentum*, entrails). The cavity containing the stomach, liver, intestines, kidneys, etc

Ab sorp'tion (*Lat. absorbere*, to swallow, or take in). The taking up of nutritive or waste matters by the blood-vessels or lymphatics.

Al-bū'men (*Lat* from *albus*, white). The name of a group of nourishing substances containing nitrogen, which resemble in nature the white of an egg.

Al-i-mĕnt'a-ry (*Lat. alimentarius*, from *alere*, to nourish). Pertaining to the nourishment of the body.

A-năt'o-my (*Gr anatemnein*, to cut up) The science which deals with the structure of living things.

An'eū-rism (*Gr. aneurisma*, a widening) A swelling or tumor due to unhealthy dilatation of an artery.

Ā-or'ta (*Lat*). The great artery arising from the left ventricle of the heart.

Ā'que-ous (*Lat. aqua*, water) Like water.

Ar'ter-y (*Gr arteria*, the windpipe) The name given to vessels which carry blood from the heart, these vessels were supposed by the old anatomists to convey only air, hence the name.

Ar-tĭc'ū-lar (*Lat articularius*). Pertaining to a joint.

Ar-tĭc-ū-la'tion (*Lat. articulatio*). The joining of bones in the skeleton.

Au'ri-cle (*Lat. auricula*, a little ear). The name given to the chambers at the base of the heart, which receive blood from the veins, because they have projections which resemble in form the ears of some quadrupeds

Au'di-to-ry (*Lat. audire*, to hear). Pertaining to the sense of hearing.

Bī'ceps (*Lat* having two heads) The name given to muscles which split at one end, so as to have there two separate attachments to the skeleton.

Bī-cŭs'pid (*Lat bis*, twice, *cuspis*, a point). The name of teeth which have two points on the crown.

Brŏn'chi-al (see *Bronchus*). The name of the branches of the windpipe inside the lungs.

Brŏn-chi'tis. Inflammation of the bronchial tubes ; a cold " on the chest."

Brŏn'chus (*Gr. bronchos*, the windpipe). The name of the two branches into which the windpipe divides in order to reach each lung.

Ca-nine' (*Lat. caninus*, pertaining to dogs). The pointed teeth, on each side of the incisors, which are very large in dogs

Căp'il-la-ry (*Lat. capillus*, hair). The name given to the smallest blood-vessels, because they are so slender

Car'di-ac (*Gr. kardia*, the heart; also the stomach) The name of the opening of the gullet into the stomach, it lies near the heart.

Car'pal (*Gr. karpos*, the wrist) The name given to the wrist-bones.

Car'ti-lage (Lat cartilago) The technical name of gristle; an elastic flexible material found in the skeleton.

Cā'se-ine (Lat. caseus, cheese). An albumen found in milk. When milk turns sour the caseine curdles, and when the whey is squeezed out of the curd, it remains as cheese.

Cĕll (Lat cella, a room or cellar) The name of the tiny microscopic elements which, with slender threads or fibres, make up most of the body. they were once believed to be little hollow chambers, hence the name. Most animal cells are not hollow.

Cĕm'-ĕnt. The substance which forms the outer part of the fang of a tooth

Cĕr-e-bĕl'lum (Lat. dim ot cerebrum, brain). The hinder and lower division of the brain The small brain

Cĕr'e-bro spi'nal. Pertaining to the brain and spinal cord.

Cĕr'e-brŭm (Lat.). The chief division of the brain. The large brain.

Chō'roid (Gr. chorion, a membrane, and eidos, form). The middle membrane or coat of the eyeball

Chȳle (Gr. chulos, juice) The digested nutritious part of the food prepared in and absorbed from the intestines

Chȳme (Gr. chumos). The name of the partly digested food which passes from the stomach to the intestine

Clāv'i-cle (Lat clavicula, a small key). The collar-bone so named because it somewhat resembles in form an ancient key.

Co-ăg-ūlā'tion (Lat coagulatio). The act of turning from a liquid to a semi-solid state. The clotting of blood.

Coc'cyx (Gr. kokkux, a cuckoo) The lowest bone of the spinal column, named from a fancied resemblance in form to the bill of a cuckoo.

Coch'le-a (Lat. cochlea, a screw). A coiled or twisted portion of the internal ear

Con'cha (Lat. a shell). The portion of the ear which projects from the side of the head

Con-gĕs'tion (Lat congestio, the act of gathering into a heap) An unhealthy accumulation of blood in any part of the body.

Con-nect'ive tissue A tough stringy material used for binding together the parts of the body.

Con-junc'ti-va (Lat. conjunctivus, serving to unite) The name of the thin membrane which lines the inner side of the eyelids and covers the front of the eyeballs.

Con-trac'tion (Lat. contractio, a drawing together). The shortening of muscles when they work.

Con-vo lu'tion (Lat. convolutus, twisted together) The winding ridges on the surface of the brain.

Cor'ne-a (Lat corneus, horny). The transparent membrane in front of the eye.

Cor'pus-cle (Lat. corpusculum, dim. of corpus, body) The name given to the minute particles which float in the blood-liquid.

Crŷs'tal-line (Gr krustallinos, ice-like, or resembling transparent crystal) The name of the lens of the eye.

Cū'ti-cle (Lat. cuticulus, dim. of cutis, skin). The outer layer of the skin; the epidermis.

De-gĕn-er-ā'tion (Lat degenerare, to grow worse; to deteriorate) A change in the structure of any organ which makes it less fit to perform its duty or function

Dĕg-lu ti'tion (Lat deglutire, to swallow down). The act or process of swallowing.

Dĕn'tine (Lat. dentis, of a tooth). The hard substance which forms most of a tooth. Ivory.

Der'mis (*G- derma*, the skin or hide) The deeper layer of the skin, containing blood-vessels.

Dī'a-phragm (*Gr. diaphragma*, a partition-wall). The muscular membrane which separates the cavity of the chest from that of the abdomen.

Dī-ar-rhœ'a (*Gr diarrein*, to flow through) An unnaturally frequent and liquid evacuation of the bowels.

Di-gĕs'tion (*Lat. digestio*, the distribution of food through the body). The process of preparing the nutritious parts of the food for absorption from the alimentary canal

Dis-lo-cā'tion (*Lat. dislocare*, to put out of place). The name of an injury to a joint, in which the bones are forced out of their sockets.

Dor'sal (*Lat. dorsum*, the back) Pertaining to the back of the body.

Dŭct (*Lat. ductus*, a leading or drawing). A tube by which fluid is conveyed from a gland.

Dys-pĕp'si-a (*Gr dūs*, ill, *pessein*, to digest). A condition of the alimentary canal in which it digests imperfectly Indigestion

En-ăm'el The smooth hard substance which covers that part of a tooth which projects beyond the gum

Ĕp-i-dĕrm'is (*Gr. epi*, upon, *derma*, skin) The outer layer of the skin The cuticle

Ĕp-i-glŏt'tis (*Gr epi*, upon, *glotta*, tongue). A cartilage at the root of the tongue which closes the opening from the throat to the larynx during swallowing.

Ĕp'i-lĕp-sy (*Gr epilepsis*, a failure or lack) A nervous disease accompanied by fits in which consciousness is lost The falling sickness.

Eū-stā'chi-an (from an Italian anatomist named Eustachi). The tube which leads from the throat to the middle ear or tympanum.

Ex-crē'tion (*Lat. excretus*, sifted out). The act of removing waste matters from the body Also any such waste matter.

Ex-pi-rā'tion (*Lat. expiro*, I emit, or breathe out). The act of expelling air from the lungs

Faū'cēs (*Lat.*) The part of the mouth which opens into the pharynx.

Fĕ'mur (*Lat.*) The thigh-bone.

Fī'bre (*Lat. fibra*, a filament). One of the slender threads of which many parts of the body are composed

Fī'brinĕ. The solid substance which forms in blood when it clots.

Fīb'ū-la (*Lat* a clasp or buckle) The outer or small bone of the leg, running from knee to ankle.

Fŏl'li-cle (*Lat folliculus*, a small bag). A little cavity or pit.

Fo-rā'men (*Lat.*) A hole or aperture

Fŭnc'tion (*Lat. functio*, a performing or executing) The special action or duty of any organ of the body.

Frŏnt'al (*Lat frons*, the forehead). The bone which supports the forehead and closes the front of the skull-chamber.

Găn'gli-on (*Gr.* a swelling) One of the smaller nerve-centres.

Găs'tric (*Gr. gaster*, the belly). Belonging to the stomach.

Glănd. An organ which forms or separates from the blood some peculiar liquid, either for use in the body (secretion), or for removal from it (excretion)

Glŏt'tis (*Gr. glotta*, the tongue). The narrow opening between the vocal cords

Hĕm'or-rhage (*Gr. haima*, blood; *regnunai*, to burst). Bleeding

He-păt'ic (*Gr. hepatikos*). Pertaining to the liver.

Hă·me-rus (Lat) The bone of the arm between shoulder and elbow

Hă'mor (Lat. moisture). The transparent liquid or semifluid substances within the eyeball.

Hȳ'gi-ēne (Gr. Hygeia, the goddess of health). That department of knowledge which deals with the preservation of health

Hȳ'oid (Gr. the letter *u,* and *eidos,* form). U-shaped. The name of the bone at the root of the tongue.

In-cī'sor (Lat. incidere, to cut into) The name of the front teeth.

In-spi-rā'tion (Lat. inspirare, to blow or breathe in or upon) The act of drawing a breath.

In-tĕs'tines (Lat. intestinus, inward). The coiled tube conveying food from the stomach. The bowels.

In-ver'te-brate. Term applied to animals having no back-bone.

In-vŏl'un-tary (Lat. in, not, *voluntarius,* acting on free choice). Performed without direction from the will, often against the will.

I'ris (Lat. the rainbow). The colored part of the eye surrounding the pupil.

Ja'gu-lar (Lat. jugulum, the hollow part of the neck above the collar-bone). The name of the chief veins of the neck.

Lăb'y-rinth (Gr. labyrinthos, a place full of intricate winding passages). The name of the inner portion of the ear.

Lăch'ry-mal (Lat lacrima, a tear). Pertaining to or conveying tears.

Lăc'te-al (Lat. lacteus, milky). The name of the lymphatics or absorbents of the small intestine During digestion they are filled with milky-looking chyle.

Lăr'ynx (Gr.) The portion of the air-passage, above the windpipe, in which voice is produced.

Lig'a-ment (Lat. ligamentum). One of the cords or bands used to bind bones together at joints.

Lum-bā'go (Lat lumbus, a loin). A painful rheumatic disease of the muscles of the small of the back

Lȳmph (Lat. lympha, water). A colorless liquid which exudes from tne blood-vessels and bathes the tissues and organs.

Lym-phăt'ic. The name of the vessels which contain lymph. The absorbents.

Mă'lar (Lat. mala, the cheek). The name of the cheek-bone.

Măl'le-ŭs (Lat. hammer). The name of the outermost bone within the middle ear.

Măm-măl'i-a (Lat mamma, a breast). The name given to the highest division of back-boned animals, because their females suckle the young.

Măs-ti-cā'tion (Lat masticatio). The act of chewing

Max-il'la (Lat the jaw). The name of the jaw-bones, upper and lower.

Me-ā'tus (Lat. a going or course). A passage or channel, as the external auditory meatus which leads from the outer to the middle part of the ear.

Me-dŭl'la ob-lon-gā'ta (Lat. the prolonged or continued marrow) The continuation of the spinal cord (*medulla spinalis*) or marrow, which enters the skull.

Mĕm'brāne (Lat. membrana, the thin skin covering the members or limbs). A thin sheet of tissue used to wrap and protect various organs, or to line cavities in the body.

Mĕt-a-car'pal (Gr meta, beyond; *karpos,* the wrist). The name of the bones between the wrist and the fingers.

Mĕt-a-tar'sal (Gr. from *meta,* beyond, and *tarsal,* which see) The name of the bones in the front part of the sole of the foot.

Mī'tral (*Lat. mitra*, a head-band). The name of the valve between the left auricle and ventricle of the heart, which has two flaps, like the mitre of a bishop.

Mō'lar (*Lat. mola*, a well). The name of the grinding-teeth.

Mō'tor (*Lat movere*, to move). Concerned in producing movement

Mū'cus (*Lat. mucus*, the secretion of the nose). A viscid liquid secreted by certain membranes within the body, named mucous membranes

Nar-cŏt'ic (*Gr. narkotikos*, from *narke*, numbness). Any substance which dulls the sensibility of the nerves, and in larger doses produces unnatural sleep.

Nā'sal (*Lat. nasus*, the nose). Pertaining to the nose; the name of the bones which support the bridge of the nose.

O-dŏn'toid (*Gr. odontos*, of a tooth; *eidos*, shape). The name of the bony peg of the second vertebra, around which the first turns.

Œ-sŏph'a-gŭs (*Gr. œsophagos*). The gullet. The tube which conveys food from the throat to the stomach

Ol-făc'to-ry (*Lat. olfacere*, to smell). Pertaining to the sense of smell.

Ŏp'tic. Pertaining or related to the sense of sight.

Or'gan (*Lat. organum*, an instrument or implement). A portion of the body having some special function or duty.

Păl-pi-tā'tion (*Lat. palpitatio*, a frequent or throbbing motion). A violent and irregular beating of the heart.

Păn'cre-as (*Gr. pan*, all; *kreas*, flesh). One of the most important glands which aid in the digestion of food. It is placed in the abdomen, just below the stomach, and pours its secretion into the upper end of the small intestine.

Pa-pĭl'la (*Lat.* a nipple or teat). The name of the small elevations found on the skin and mucous membranes.

Pa-răl'y-sis (*Gr. paraluein*, to set free or separate). Loss of function, especially of motion or feeling. Palsy.

Pa-rī'e-tal (*Lat. paries*, the wall of a house). The name of the bones on the top of the skull

Pa-tĕl'la (*Lat*). The knee-cap or knee-pan.

Pĕl'vis (*Lat.* a basin). The bony ring, made of sacrum, coccyx, and the two hip-bones, which surrounds the lower part of the abdomen.

Pĕr-i-car'di-ŭm (*Gr. peri*, around; *kardia*, the heart). The membranous sac which encloses the heart.

Pĕr-i-ŏs'te-um (*Gr. peri*, around; *osteon*, a bone). A fibrous membrane which surrounds the bones.

Pha-lăn'gēs (*Gr. phalanx*, a body of soldiers closely arranged in ranks and files). The bones of the fingers and toes.

Phăr'ynx (*Gr.* the throat). The cavity into which the nose and mouth open, and from which the gullet proceeds.

Phys-i-ŏl'o-gy (*Gr. physis*, nature; *logos*, a discourse). The science which treats of the functions or uses of the different parts of animals and plants.

Plăs'ma (*Gr.* anything formed or moulded). The liquid part of the blood.

Pŭl'mo-na-ry (*Lat. pulmonis*, of a lung). Pertaining to the lungs.

Pȳ-lō'rus (*Gr. pyloros*, a door-keeper). The opening from the stomach into the small intestine.

Rā'di-ŭs (*Lat.*). The outer of the two bones running from the elbow to the wrist.

Rē'flex (*Lat reflexus*, turned back). The name given to involuntary movements produced by an excitation travelling along a sensory to a centre, where it is turned back or reflected along motor nerves.

Rĕ'nal (*Lat renes*, the kidneys).   Pertaining to the kidneys.

Rĕt'in-a (*Lat. rete*, a net).   The transparent nervous membrane which forms the inner coat of the eyeball.

Sā'crum (*Lat.* sacred)   The large bone near the lower end of the spine, having the hip-bones attached to its sides

Sa-lī'va (*Lat.*).   The liquid which moistens the mouth, and aids in swallowing and digesting

Sa-phē'nous (*Gr saphenes*, manifest).   The name of a large vein which lies just under the skin of the leg

Scăp'a-la (*Lat*)   The shoulder-blade

Scle-rŏt'ic (*Gr. skleros*, hard, tough).   The tough outer coat of the eyeball.

Se-bā'ceous (*Lat sebum*, tallow)   The name of the oil-glands of the skin

Se-crē'tion (*Lat. secretio*, a separating).   The preparation, from the blood, by glands, of peculiar liquids.

Sem-i-lū'nar (*Lat. semi*, half; *luna*, mooned)   Shaped like a half-moon.

Sen-sā'tion (*Lat sensus*, feeling)   Any kind of feeling, as hunger or hearing.

Sē-rum (*Lat.* whey).   The liquid part which separates from the clot, when blood coagulates.

Skĕl'e-ton (*Gr.* dried up).   The bones and other supporting parts of the body, as gristles and connective tissue.

Sphē'noid (*Gr. sphen*, a wedge, *eidos*, form).   The name of one of the bones on the under side of the skull.

Stā'pēs (*Lat.* a stirrup).   The name of the innermost bone of the middle ear, which has the form of a stirrup

Ster'num (*Gr sternon*, the chest).   The breast-bone.

Stim'ū-lant (*Lat. stimulare*, to goad or stir up).   Any substance which excites some organ of the body to do extra work, without proportionately nourishing it

Sū-dor-ip'a-rous (*Lat. sudor*, sweat; *parare*, to prepare).   The name of the glands of the skin which secrete sweat or perspiration.

Sut'ūre (*Lat. sutura*, a seam).   The union of certain bones of the skull by the interlocking of jagged edges

Syn-ōv'i-al (*Gr. syn*, with; *ŏŏn*, an egg).   The liquid which lubricates the joints, joint-oil.   So called from its resemblance to the white of a raw egg

Tar'sal (*Gr. tarsos*, a broad, flat surface, hence the sole of the foot).   The name of the bones below the ankle-joint.

Tĕm'po-ral (*Lat. tempora*, the temples).   The name of the skull-bones which support the temples, and contain the inner parts of the ear.

Tĕn'don (*Lat tendere*, to stretch).   The cords which attach muscles to bones.

Thō'rax (*Gr.* a breast-plate).   The chest.   The upper part of the trunk of the body.

Tĭb-i-a (*Lat.*)   The shin-bone.

Tis'sue (*Lat texere*, to weave).   The name given to each of the materials used in the construction of the body, as muscular tissue, nervous tissue, bony tissue, etc.

Trā'che-a (*Gr. trachus*, rough).   The windpipe.

Tri-cŭs'pid (*Lat. tris*, three times; *cuspis*, a point)   Having three points   The name of the valve between the right auricle and ventricle of the heart.

Tÿm'pa-num (*Lat.* a drum).   The middle or drum chamber of the ear.

Ul'na (*Lat.*).   One of the two bones passing from elbow to wrist.   It lies on the inner or little-finger side.

U'vu-la (Lat a little grape). The fleshy conical body which hangs down from the lower border of the soft palate.

Văr'icōse (Lat. varix). The term applied to an unhealthily distended vein.

Văs'cŭ-lar (Lat. vasculum, a little vessel) Pertaining to or possessing blood- or lymph-vessels

Věn'tral (Lat. venter, the belly) Pertaining to the front or belly side of the body.

Věn'tri cle (Lat ventriculus the belly). A small cavity, as the ventricles of the heart. Also applied to cavities within the brain.

Ver-te'bra (Lat. from *vertere,* to turn). The name of each of the bones of the spinal column

Věs'ti-bale (Lat. a fore-court or entry to a house). A part of the inner ear from which the other parts open.

Vil'lus (pl. *vil'li, Lat.* shaggy hair). The name of the minute hair-like projections of the mucous membrane of the small intestine

Vit're-oŭs (Lat. vitreus, glassy) One of the substances within the eyeball, which guide rays of light to the retina.

Vŏl'un-ta-ry (Lat. voluntarius). Applied to actions performed in obedience to the will.

INDEX.

SCIENCE TEXT BOOKS

PUBLISHED BY

HENRY HOLT & CO.

ASTRONOMY.

BALL'S ASTRONOMY. By R. S. BALL, LL.D., F.R.S., Astronomer Royal for Ireland. Specially revised by SIMON NEWCOMB, LL D., Professor of Mathematics, U. S. Navy. 16mo 60 cents.

CHAMPLIN'S (JOHN D.) THE YOUNG FOLKS' ASTRONOMY. *Very simple and elementary.* 16mo. Illustrated. 60 cents.

NEWCOMB AND HOLDEN'S ASTRONOMY FOR SCHOOLS AND COLLEGES. By SIMON NEWCOMB, LL.D., Professor of Mathematics, U. S Navy, and EDWARD S. HOLDEN, M.A., Director of the Washburn Observatory, University of Wisconsin. With numerous illustrations. Third edition, revised and partly rewritten. Large 12mo. (American Science Series). $2 50.

THE SAME. BRIEFER COURSE. 12mo. $1.40.

BOTANY.

BESSEY'S BOTANY. For Students and General Readers. By C. E. BESSEY, Professor of Botany in the Iowa Agricultural College. With over 500 illustrations. Large 12mo (American Science Series). $2.75.

THE SAME. BRIEFER COURSE. 12mo. $1.35.

KOEHLER'S PRACTICAL BOTANY. Structural and Systematic. The latter portion being an analytical key to the wild-flowering Plants, Trees, Shrubs, &c.; Ordinary Herbs, Sedges and Grasses of the Northern and Middle United States, East of the Mississippi. By AUGUST KOEHLER, M.D, Professor of Botany in the College of Pharmacy of the City of New York. With numerous illustrations by the author. Large 12mo. $2.50.

MACLOSKIE'S ELEMENTARY BOTANY. With Student's Guide to the Examination and Description of Plants. By GEORGE MACLOSKIE, D.Sc., LL.D., Professor in Princeton College, N. J. 12mo. $1.60.

McNAB'S BOTANY. Outlines of Morphology, Physiology, and Classification of Plants. By WM. RAMSAY McNAB, Professor of Botany in Royal College of Science for Ireland. Revised for American Students by Prof. C. E. BESSEY. 16mo $1.00.

PHYSIOLOGY.

MARTIN'S "HUMAN BODY." An account of its structure and activities, and the conditions of its healthy working By H. NEWELL MARTIN, Professor of Biology in the Johns Hopkins University Large 12mo. (American Science Series). With Appendix on Reproduction and Development. $2 75. Copies without the Appendix will be sent when specially ordered.

THE SAME. BRIEFER COURSE, with special chapter on the action of Alcohol and other stimulants and narcotics. 12mo. $1.50.

THE SAME. ELEMENTARY COURSE, with special reference to the effects of Alcoholic and other stimulants, and of narcotics. 12mo. 90 cents.

POLITICAL ECONOMY.

ROSCHER'S POLITICAL ECONOMY. By WILLIAM ROSCHER, Professor of Political Economy at the University of Leipzig. With additional chapters furnished by the author, for this First English and American edition, on Paper Money, International Trade and the Protective System, &c.; and a preliminary Essay on the Historical Method of Political Economy (from the French, by L. WOLOWSKI). The whole Translated by JOHN J. LALOR, A.M. 2 vols., 8vo. $7.00.

SUMNER'S (W. G.) PROBLEMS IN POLITICAL ECONOMY. By WILLIAM GRAHAM SUMNER, Professor of Political and Social Science in Yale College. 16mo $1.25.

WALKER'S POLITICAL ECONOMY. By FRANCIS A. WALKER, President of the Massachusetts Institute of Technology, late Superintendent of the Census. Large 12mo. (American Science Series). $2.25.

THE SAME. BRIEFER COURSE 12mo. $1.50.

ZOOLOGY.

MACALISTER'S ZOOLOGY. Zoology of the Invertebrate Animals. By ALEX. MACALISTER, M.D., Professor of Zoology and Comparative Anatomy in the University of Dublin. Specially revised for America by A. S PACKARD, JR., M.D., Professor of Zoology and Geology in Brown University. 16mo. $1.00.

PACKARD'S ZOOLOGY. For Students and General Readers. By A. S PACKARD, JR, M D., PH.D, Professor of Zoology and Geology in Brown University. With over 500 illustrations. Large 12mo. (American Science Series). $3.00.

THE SAME. BRIEFER COURSE. 12mo. $1 40.

☞ A specimen copy of any of the foregoing. except *Roscher's Political Economy*, sent post-paid to a Teacher upon receipt of half the retail price.

HENRY HOLT & CO., Publishers, New York,

Milton Keynes UK
Ingram Content Group UK Ltd.
UKHW022241011223
433582UK00006B/195